The Universe-an Irrelevant Illusion

Mike Joslin

Published by New Generation Publishing in 2024

First Edition

ISBN: 978-1-83563-409-7

www.newgeneration-publishing.com

New Generation Publishing

Acknowledgements

I am indebted to my wife Kath, for all of the love, patience and support she has shown me on our energetic and busy voyage through life together. Thanks also to my extended family for their love and kindness at difficult times.

Contents

The Universe-Fact or Fantasy

There is irrefutable evidence that what we observe in the Universe is mostly a mirage. I hope we can all move away from the Big Bang Theory and open our eyes to a much more interesting interpretation. We desperately need to look for truth in this troubled era.

I'm sure Albert Einstein would laugh in heaven if he read my paper since he always considered imagination more valuable than knowledge. The great man is credited with having made the following remarks. Undoubtedly, he possessed an incredible brain and exhibited a wonderful sense of humour.

> *"A man should look for what is, and not for what he thinks should be".*
> *"In the middle of difficulty lies opportunity."*
> *"A human being is part of a whole called by us the universe".*
> *"The important thing is to not stop questioning. Curiosity has its own reason for existing."*
> *"Concern for man and his fate must always form the chief interest of all technical endeavours. Never forget this in the midst of your diagrams and equations."*

"There are only two ways to live your life. One is as though nothing is a miracle. The other is as though everything is a miracle."
"Once you stop learning, you start dying".
"It has become appallingly obvious that our technology has exceeded our humanity."

Until we all move on to a much more plausible alternative explanation of the Universe, we will continue to endlessly invent new ways to justify the inexplicable. We should dispose of the Big Bang Theory and 'open the skies' to a much more exciting and tangible possibility.

Scientists have assumed that innumerable stellar objects are so distant from us that they can only have got there by way of an initial explosion so enormous that matter accelerated away from its centre at a greater velocity than that of light. However, would we see stars if they were distancing themselves faster than the speed of light since this assumption is necessary to support the Big Bang idea.

This was reached on the basis that the 'red shift of an object's light (photons) occurs when it is moving away from an observer. This is described as the 'Doppler Effect', the change in frequency of a light/sound wave in relation to an observer who is moving relative to the source of the wave. Although the Doppler Effect is quite authentic in relation to how sound is affected by the velocity of its source, I find it difficult to apply this to spacial events. Nevertheless, red shift has been used as 'proof' that a 'Big Bang' occurred and that the Universe is therefore expanding at an accelerating rate greater than the speed of light. This theory is in contrast to Einstein's earlier opinion that nothing can exceed the speed of light. I dealt with that subject in my book, "Einstein's $E=mc^2$ unravelled". Suffice to say however that if the Universe did expand at faster than the speed of light with a 'Big Bang', where did it get the energy from in the first place?

I intend therefore to keep things simple. Astronomers continue to use 'red shift' as a way of determining how far away from Earth a stellar object is and correspondingly therefore, its rate of departure/acceleration from the Universe's centre; where we are! Why should we be at the centre?

This article was published in the Guardian newspaper on the 5th April 2024 under the heading,, "*3D cosmic map raises question over future of universe and dark energy, scientists say........Researchers said that by using this map, they have been able to measure how fast the universe has been expanding at different times in the past with unprecedented accuracy.....the results confirm that the expansion of the universe is speeding up.......*"

I hope to describe the Universe more realistically. Light (photons), travelling through a complete vacuum neither gains nor loses energy without cause. We have presumed that a star's 'red shift' is indicative of its <u>radial</u> distance from us (and acceleration away) when in my opinion it is actually a measure of the energy lost by light from a radiant body such as a star en route to an observer. I will explain that in due course.

Unfortunately, the Big Bang theory is now universally accepted by the scientific community as proof that the Universe is still expanding in every direction from its centre to its current 'radius' of about 45 billion light-years and is still expanding at an accelerating rate. To support that idea, scientists have been obliged to search for the source of energy to justify this interpretation and settled on the Boson, a mystical particle evading firm proof of its existence. So far, it appears to be reluctant to be indisputably identified and in my eyes therefore, open to conjecture. There are some alternative explanations which have been largely ignored but it seems that having decided on the Big Bang, its enormous source of energy had to be found!

If light from a star travels to us here on Earth through the complete vacuum of outer space without countering any gravitational obstacles, it cannot lose its energy. Its 'red shift' will therefore be unaffected. The following article metaphorically 'cast a huge spanner in the 'Big Bang' 'works'.

From the web, "Sir Arthur Eddington led an expedition to photograph the **1919 Total Eclipse of the Sun**. Photographs revealed stars whose light had passed near to the Sun. Their positions showed that their light had been bent exactly as Einstein had predicted." The following explains the subject admirably.

Light reaching Earth from stars billions of light-years away is bent by our sun's gravity. The observed bending of light from stars during the experiment in 1919 was due entirely to the sun's gravity and was measurable <u>after only 8 light-minutes</u> of 'travel-time' following its 'brushing close' to our sun, before continuing to its final destination here on Earth. Importantly, it provided an answer as to how light energy is depleted by gravitational bodies on its way to an observer. Light's loss of energy due to its being gravitationally bent consumes some of its energy thus affecting its red shift. Once a stars' light is bent by a gravitational force(s), it becomes virtually impossible to review its prior history since we can only view its apparent position in the sky.

Let us suppose that during the 1919 experiment, some 'sun-bent' starlight just missed the Earth and continued on its gravity-bent path for a further 10 light-years before reaching another planet whose imaginary scientists viewed it using a similar telescope to ours. By then, the effect of the original deflection of the stars' light by our sun would have been magnified with distance travelled.

Please allow me to state the most important news in my entire book.

Due to the proven gravitational bending of starlight, our sun constantly 'gives birth' to an infinite number of phantom 'replicas'. This phenomenon is universal. Light reaching every star from every other one within mutual visible range will behave thus.

Of course, many of the virtual 'copies' generated by stars will in turn be replicated providing us with even more 'fairy lights' to fill the sky with!
Light from any radiant body in the Universe exhibiting red-shift should be dispatched immediately to the dustbin to join the Big Bang Theory. The only authentic ones are those without it.

A friend mentioned to me that he had found it difficult to picture how two separate images from one star could be viewed simultaneously. The simple illustration below illustrates how this occurs trillions of times in the Universe. Some light from the star on the left is bent by the gravitational effect of the one on the right. Some light travels directly to an observer unaffected by grav ty. The parcel of light which travels furthest to the same observer travels farthest en route <u>loses energy due to gravitational effects and is thus red-shifted</u>.

The differing 'angles of incidence' of 'parcels of light' from one star passing close enough to be bent by the gravity field of all the other stars or galaxies in their mutually visible Universe, create numerous 'duplicate' images marked in green

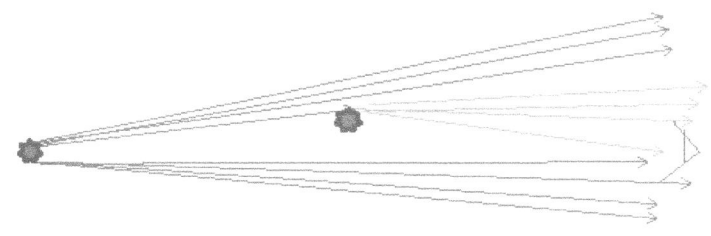

Ethan Siegel wrote: *"What happens to light when it passes near a large mass? Does it simply continue in a straight line, undeflected from its original path? Does it experience a force owing to the gravitational effects of the matter nearby? And if so, what is the magnitude of the force it experiences?*

"These questions cut to the very heart of how gravity works. In 2019, the 100th anniversary of General Relativity was confirmed. Two independent teams undertook a successful expedition to measure the positions of stars near

the limb of the Sun during the total solar eclipse of May 29, 1919. Through the highest-quality observations that technology permitted at the time, they determined whether that distant starlight was bent by the Sun's gravity, and by how much. It was a result that shocked many, but Einstein already knew what the answer would be. Here's how even before Einstein had worked the theory out, he knew that light must be bent by masses." (In fact we know now that the infinite gravity of a 'black hole' can 'swallow' light without trace!)

"The conclusion to Einstein's thought experiment was irrefutable. Whatever the gravitational effects are at a certain location in space — whatever accelerations they induce — they will affect light as well. Just as surely as accelerating your elevator with thrust will cause a light ray to deflect; accelerating it by having it in the proximity of a gravitational mass will cause that same deflection.

"Therefore, Einstein reasoned, it would not only be possible to predict that light rays cannot travel along a straight path when they're in a gravitational field, but the magnitude of deflection could be calculated simply by

knowing what the strength of the gravitational effects in the vicinity of that mass were."

The experiment carried out 105 years ago proved conclusively that light from several stars was bent by our Sun. This happened literally 'on our doorstep,' and was measurable, so we shouldn't ignore its implications. Since that time, we appear to have ignored the fact that red shift measurements are explicably due to the gravitational effects of other bodies. This is a vastly more coherent theory than ascribing the cause of red-shift to celestial bodies' acceleration away from us!

We should have used our imaginations as Einstein encouraged us to do. When we look at stars whose light indicates a 'red shift', shouldn't we be relating that to the likelihood that the path of its light has been affected by passing close to other stars? This could lengthen its light's journey compared to observers on other planets the same radial distance from the star as Earth but whose light had not been affected in that way. The shortest distance between two points is a straight line! The imaginary scientists 10 light-years on from Earth would have no way of knowing this.

The 'Higgs Boson' was subsequently 'invented' as a means to justify the energy requirement of the supposed 'Big Bang' as I described. Despite spending £100's of millions building the Hadron Collider in an attempt to isolate and confirm the presence of Bosons, it remains questionable. Surely, we should have opened our minds to completely new ways of analysing our Universe. It seems fitting at this point to repeat some of Einstein's views:

"The important thing is to not stop questioning. Curiosity has its own reason for existing."

"Concern for man and his fate must always form the chief interest of all technical endeavours. Never forget this in the midst of your diagrams and equations."

Lght reaching us from stars provides us only with apparent indications of their locations and distance from us and that red-shift can provide us with clear-cut evidence of light's loss of energy. We knew in 1919 that light from stars loses energy when it passes close to gravitational bodies and gets bent in the process.

Far too many assumptions have been made on the basis that 'red shift provides us with evidence of the expanding distance and thus the acceleration of objects away from us. Swedish astronomer <u>Knut Lundmark</u> was the first person to find observational evidence for expansion in 1924. According to Ian Steer of the NASA/IPAC Extragalactic Database of Galaxy Distances, "Lundmark's extragalactic distance estimates were far more accurate than Hubble's, consistent with an expansion rate (Hubble constant) that was within 1% of the best measurements today. Here is a quotation from Wikipedia:

> *"In physics, redshift implies an increase in the wavelength, and correspondingly, a decrease in the frequency and photon energy of light. The opposite change, a decrease in wavelength and increase in frequency and energy, is known as a blue-shift, or negative redshift. The terms derive from the colours red and blue which form the extremes of the visible light spectrum. The main causes of electromagnetic redshift in astronomy and cosmology are the relative motions of radiation sources, which give rise to the relativistic Doppler effect, and gravitational potentials, which gravitationally redshift escaping radiation. All sufficiently distant light sources show cosmological redshift corresponding to recessional speeds proportional to their distances from Earth, a fact known as Hubble's law that implies the universe is expanding."*

I cannot agree. This article ignores the certainty that stars' loss of energy is far more likely to be caused by close encounters with gravitational bodies. Stars' light can be bent invisibly to an observer at any time en route as depicted earlier and its red-shift caused by a loss of energy due to those gravitational encounters and will result in lengthier journeys. Light does not lose energy in a complete vacuum. The 1919 experiment demonstrated conclusively that the bending/deflection of a star's light from its original path by a gravitational body was evident after travelling only 8 light-minutes to observers on Earth and provided them with only its apparent 'starting point! It cannot conceivably provide him with evidence of either its range or actual location. On the contrary, light reaching us from a star with no red-shift is reliably indicative of its <u>actual</u> position in the sky! Bent starlight cannot indicate either a star's actual position or its range. This process can cause stars to appear as if they have a twin!

The bending of light from radiant bodies by their gravitational encounters with other stars en route to Earth will lengthen the distance it travels to us. The sum of the lengths of two sides of a triangle will always be greater than the length of the third! Its journey to an observer will of course take more time. By ignoring these gravitational effects, we have to depend on the theory that the universe is expanding exponentially. The energy-loss of light reaching us from stars and being bent gravitationally en route is far more common that we believe. Surely, red shift cannot be interpreted as both the rate of acceleration of a body away from an observer and its loss of energy due to gravitational effects.

According to the European Space Agency, *"For the Universe, the galaxies are our small representative volumes, and there are something like 10^{11} to 10^{12} stars in our Galaxy, and there are perhaps something like 10^{11} or 10^{12} galaxies. With this simple calculation you get something like 10^{22} to 10^{24} stars in the Universe."*
(10,000,000,000,000,000,000,000,000!) *"This is only a rough number, as obviously not all galaxies are the same, just like on a beach the depth of sand will not be the same in different places"*.

We seem 'blind' to the fact that 'duplicates' and even 'triplicates' and 'quadruplets' of stars with apparently older and younger histories surround us everywhere. Every image of stars/galaxies in the night sky that exhibit 'red shift' should be viewed with suspicion since it is almost certain that it will have been caused by one or more gravitational encounters and thus falsely representative of their respective and actual locations. The light from many will have experienced so many deflections that it will have faded completely.

Radiant bodies can be delineated only by their respective 'red shift' measurements until they eventually lose all of their energy and 'cease to exist' to an observer. 'Overtaking' could occur when a star's light takes a more direct route to an observer than that of adjacent stars adding to the complexities confronting our scientists! A simple analogy would be what occurs when a wave of water reaches a shoreline. Its height would eventually lessen to a trickle as its energy is dissipated completely and ending motionless on a beach. A 'twin' wave could encounter a rock en route losing some of its energy prematurely.

The number of 'false' images in the visible Universe may have reached its maximum figure. This depends of course on the power and range of our telescopes and what can only be described as a Universe in equilibrium. Already, there is a certain balance between the light from stars surviving total loss of their light energy ('disappearing completely' through multiple light-bending encounters) and those in various visible states.

Stars are spherical. This means that every star in the viewable universe emits light in every conceivable direction. If there were only two stars in the Universe within visible range of each other (an assumed 45 billion light years), light from each of them would reach the other and be bent by their respective gravitational effects. As proven 105 years ago, new images would be created. This isn't a problem when the number of stars is small. The images generated in the gravitational light-bending process are merely 'virtual' and are not real! It's like viewing a candle in a 'hall of mirrors'. When a star's light moves through densely-packed clusters of other gravitational bodies, cumulative red-shifting of their light will occur and their light will be dissipated creating innumerable 'false images'.

The vast majority of 'stars' we see in the sky constitute 'false images'. Surely, light from an estimated $10^{22-}24$ cannot reach us without making numerous gravitational encounters en route.

Wikipedia and other sources tell us that the Universe's visible diameter is approximately 90 billion light-years. We need only use our imaginations therefore to picture light's journey to Earth from the Universe's farthest visible entities. During its journey to us, stars' light will have been subjected to being bent by numerous other gravitational bodies it encounters, creating the illusion of a sky actually much more 'crowded' than it really is. Therefore, using the formula $V = 4/3\pi r^3$ (where r = 45 billion light years) and the stated 10^{22} stars in our visible universe (ignoring about a trillion galaxies), gives us the approximate volume of our visible universe as 381,750 billion cubic light-years. The average volume per star therefore is only a few cubic light years!

By now, Einstein will be laughing his head off. What scientists have been claiming for decades that everything kicked off with a Big Bang appears to be impossible! The light from any star in our entire visible Universe would have great difficulty reaching any other part of it in visible range without being bent by gravity at least once and probably many times en route to an infinite number of viewers.

How on Earth can 10^{11} stars and 2 trillion galaxies be packed so tightly together without interfering with one another's light and as a consequence, producing an infinite number of images? A huge proportion of them don't exist; they are merely virtual realities! They represent *apparent positions* of stars in the Universe in various *apparent ages* of development/decay. We mustn't forget that light reaching us from the universe's currently <u>presumed</u> visible radius will have left stars billions of light-years ago so whatever we view out there will be billions of light-years out-of-date!

Light is emitted from every luminous body in the Universe in every conceivable direction from its entire spherical surface. If only two stars existed in our 'visible' universe, each would bend the light from the other regardless of its destination creating myriads of false images: once a certain concentration of stars is reached, the number of 'false images' accumulates exponentially until we are flooded with them!

Because we are fixated on the Big Bang Theory, we have allowed ourselves to continually invent equations to support that theory. The Boson particle was 'created' mathematically to satisfy the conundrum of where the energy came from to support the theory; a source of the infinite amount of energy needed to continuously and exponentially propel trillions of stars into infinity. I believe we are forgetting what we learned a century ago that stars bend light from other stars. How many of these images are 'real' and how many are just 'copies' or virtual images of stars whose light has been bent en route to us?

Apparently, two stars shine in the evening sky. They are Pollux and Castor in the constellation Gemini the Twins. Pollux is slightly brighter than Castor. It shines with a golden glow while Castor appears whiter. Pollux is the 18th brightest star in Earth's night sky.

This could be an illustration of how a star's apparent replica/'twin' is reproduced when its light is deflected on its way to Earth. We have finished up with two views, Pollux and Castor. It's obvious which one is real since it can only be the brightest, Pollux since its journey to us here on Earth is the shortest bearing out my theory precisely. Of course, if light from that star also shows red shift, that version could also be 'phony'! The lengthened journey that the twin's light has taken is probably the reason why some of its energy has been consumed en route. This process is recurring trillions of times around us and we appear to be unable to entertain the idea!

Light reaching Earth from celestial bodies up to 45 billion light-years distance (the currently assumed radius of the visible universe) cannot conceivably reach us here on Earth without being bent en route by the gravitational effects of not only our own sun but countless other stars, galaxies and gravitational bodies encountered en route. Starlight can suffer from multiple bending occurrences creating 'red shifted' images which scientists have construed as stars' radial distances from Earth when they are anything but! The light from any star is radiated in every direction. Only a fraction of its total light emitted will reach Earth. Even if we were to equip our scientists with a telescope equipped with a lens the diameter of our planet, the amount of a star's light reaching it would still be only a tiny fraction of its total light radiated. The rest of it would be transmitted in every other direction and vulnerable therefore to greater or lesser 'bending' encounters in their labyrinthine and respective futures. What we view in our night sky is just a virtual reality! The more powerful our telescopes are, the greater the number of false images we will reveal.

The assumption that the 'red-shift' of light measurement determines the radial distances of light-emitting bodies is false. Given the facts outlined above and based entirely on proven knowledge dating back 105 years, we are currently completely misinterpreting red-shift as a measure of distance travelled/acceleration away from us. This is like attempting to measure the length of a journey made by a ship, an aeroplane or a car by measuring its fuel consumption en route? That would of course, be ignoring the effects of crosswinds, headwinds, tailwinds, tides and ocean currents which could either reduce or increase their fuel/energy consumptions. Would we measure the distance between two points by reading the mileometer in our cars since that is what we are doing!

We should never use the amount of a star's red shift as indicative of its distance. If anything, it will only indicate that it has lost some of its energy due to the combined gravitational effects of other celestial bodies. This is not the same as establishing a 'gravitational constant'! Red-shift is indicative of light reaching an observer after an indeterminable number of encounters with gravitational bodies en route. The direction it might appear to come from would bear little actual relationship to its origin. Big Bang theorists have been using the red-shift measurement of a star's light as a measurement of its radial distance when it only depicts its light's total distance travelled.

The continued use of red-shift and brightness methods to determine actual cosmic distances is absurd and invites further scrutiny. When one bases a theory on only one rule without continued corroboration, you are asking for trouble. Having assumed for a considerable time that the Universe is eternal and infinite, everything changed in 1927 and the science books had to be rewritten to new ways of thinking. The 'Big Bang' Theory took over and when distances much greater than the supposed starting point of the

Universe 13.8 billion years were *indicated*, this was overcome by inventing the term 'inflation'.

Now we constantly find ways to support the Big Bang theory ruling out common sense. This is where Nobel Prizes fail; they cement ideas deserving further challenge into cast iron respectability! Inflation needs energy so the 'Boson' was invented and remains a theoretical particle after an expenditure on research worldwide of 100's of £millions..

Light is bent when it passes large gravitational bodies in much the same way as asteroids or space exploration vessels bend going past a planet for instance. Over vast distances therefore, light will be affected by innumerable gravitational effects, some of which will be cumulative and others which will possibly negate the effects of previous ones. We must accept therefore that in most cases if not all of them, we will remain ignorant of those changes of route. Every deflection from a straight path regardless of the direction and deviations which it takes, will increase the distance travelled by light, consume it energy and thus enlarge its red shift. It is far more logical to assume therefore that light from almost all celestial objects will have been affected to greater or lesser extents by gravitational effects.

Light from stars with identical red-shift measurements could actually appear to be in almost the same place with similar red-shift measurements looking like 'twins'! We have constructed a completely false picture for ourselves of the Cosmos. The *actual positions* and *states* of celestial objects from an observer are currently impossible to calculate. We continue to believe in a Big Bang Theory entirely determined by red shift and other light decay measurements. We calculate by these methods that objects are certain distances away from us when in fact they may have been much closer when the light we see from it was emitted. This is like estimating the distance between London and Edinburgh by measuring a vehicle's fuel consumption, m.p.g., tyre wear and driver tiredness and entirely ignoring the route its driver took, its deviations, obstructions, headwinds or meal breaks!

On this basis, we cannot take anything for granted about the universe since our view and any measurements we make will only be relative to our position in relation to all other entities/objects in the past. The truth is that beyond certain distances, we cannot rely on anything, particularly red shift!

Gravitational waves and pretty drawings of space/time attempting to demonstrate how space is distorted only complicate the simple fact that the route taken by light emitted from celestial bodies is almost unavoidably distorted by the gravitational effects of all other entities it skirts. The route taken by light from other celestial bodies at vast distances can never be assumed. Twin stars *actually* equidistant from us may be revealed to us as having quite different ages based on the respective distances their light travels before reaching us and thus their respective red-shift measurements!

It's my understanding that Hubble played a huge part in establishing a century ago the fact that light is bent by gravity. I quote below from an article on Wikipedia.

> **Hubble's law**, also known as the **Hubble–Lemaître law**,[1] is the observation in physical cosmology that galaxies are moving away from Earth at speeds proportional to their distance. In other words, the farther they are, the faster they are moving away from Earth. The velocity of the galaxies has been determined by their redshift, a shift of the light they emit toward the red end of the visible spectrum. The discovery of Hubble's law is attributed to Edwin Hubble's work published in 1929.[2]
>
> Hubble's law is considered the first observational basis for the expansion of the universe, and today it serves as one of the pieces of evidence most often cited in support of the Big Bang model.[3][4]

If I was a cosmologist, I would try to establish a 'network of actuality' of the Universe by progressively eliminating stars that are practically invisible (maximum red-shift) and proceeding until the stars and galaxies in the sky are down to only the brightest (zero red shift). I think a decent computer would arrive at an almost actual state of the Universe containing a fraction of the 20^{24} currently-proposed stars and galaxies'! There would be relatively few left in the sky!

I'll close with part of a rather wonderfully synchronistic and supportive article copied from cs.unc.edu (The College of Arts and Sciences) under the heading "Red Shift Riddles":

> *"Several ways can be conceived to explain this quantization. As noted earlier, a galaxys' redshift may not be a Doppler shift, it is the currently commonly accepted interpretation of the red shift, but there can be and are other interpretations. A galaxy's' redshift may be a fundamental property of the galaxy. Each may have a specific state governed by laws, analogues to those in quantum mechanics that specify which energy states atoms may occupy. Since there is relatively little blurring on the quantization between galaxies, any real motions would have to be small in this model. Galaxies would not move away from one another; the universe would be static instead of expanding."*

Nature Philosophy & Science

The calculation for determining the kinetic energy of an object in motion is e = mv2. Einstein's equation E = mc2 treats velocity c as a constant (the speed of light) and thus m = E/c2. For any value attributed to c, it can be seen that e and m are reciprocals; E = 1/m, m = E/1. If mass becomes infinite, energy must deplete to zero. It's assumed perhaps incorrectly, that this relationship is what 'kick-started' our Universe with a 'Big Bang'. Perhaps the Universe was there in the first place as I will explain later!

There's no reason why all of the matter/energy in the Cosmos should have been in one place 13.8 billion years ago. It's perfectly possible that our bit of the Universe just happens to be much younger (or older) than much of what we can see all around us. Try thinking about the Cosmos as an infinite quantity of energy and matter which includes an infinite number of entities including 'black holes'. There are thousands of these just in our galaxy alone and a colossal number of them to be seen if we look out as far as we can see, a mere 45 billion light years. Or should I say 'were' out there!

A different way of looking at infinity is as a relative rather than a quantitative value. Beyond a certain ratio, surely all values are infinite. Attempting to define the actual amount of matter in the Cosmos is difficult to say the least; there are too many uncertainties in our current methods of calculation. As the equations above demonstrate, energy and mass are reciprocals. Looking for ways to define the total matter in the Cosmos therefore, is impracticable if one ignores the fact that much of it is invisible and possibly, immeasurable!

Taking only the matter which can be observed and assuming a quantity of 'dark energy' is therefore misleading if you don't know the ratio of matter to energy. If I'm right, since the Cosmos operates as an entropic 'closed system' it will always maintain an overall balance or equilibrium. According to Wikipedia, 'The law of conservation of mass or principle of mass conservation states that for any system closed to all transfers of matter and energy, the mass of the system must remain constant over time, as a (closed Sic) system's mass cannot change, so quantity cannot be added nor removed. Hence, the quantity of mass is conserved over time.'

And 'In physics, the law of conservation of energy states that the total energy of an isolated system remains constant, it is said to be conserved over time. This law means that energy can neither be created nor destroyed; rather, it can only be transformed or transferred from one form to another.' There will be huge variations across the Cosmos in the ratio of energy to matter.

Calculations based on Earth's 'local' conditions will only confuse us. Let's assume for the purpose of my 'mind experiment', that there is an infinite amount of matter/energy in the Cosmos and put the subject of 'space' to one side for the time being. When thinking about the Cosmos as a whole, 'space' is not only irrelevant but non-existent according to Quantum Theory. Currently, there are huge problems concerning how to mathematically compromise space-time Relativity with Quantum Theory. I argue that in the Cosmos, Thermodynamic space/time relativity and Quantum Theory co-exist. Let's call it my 'Two-State Theory'.

Many scientists presume that the Universe was created following a 'Big Bang' event when an inflation of energy into space took place. They usually talk in singular terms of 'the Big Bang' rather than perhaps an endless series of continuous conversions of matter into energy to maintain overall equilibrium. Suggesting a 'Big Bang' accounted for the creation of the Universe creates a frustrating 'enigma'.

It's assumed that The 'Big Bang' occurred when matter had somehow collapsed into a tiny volume and turned into energy like an atomic bomb. The Universe we experience and measure is governed by the 'state' of Thermodynamic Laws and most of what we perceive around us has evolved in ways which respect them. Thermodynamics directly determines the behaviour of subatomic particles when they are constituents of matter in one 'state' but only indirectly when they are 'free'. The latter 'state' is Quantum.

Each 'state' operates independently of the other but because the Cosmos acts as a 'closed' Entropic system, an overall equilibrium ratio of matter/energy must be maintained as I explained above. In my opinion, the Cosmos is the one and only example of perpetual motion and theoretically at least, the only truly 'closed' state there is. No other state can be said to fulfil the essential requirement of perpetual motion which is a system in constant change entirely closed to all outside influences, since only the Universe exists!

Attempts to interpret the nature of our Universe have largely been made on the basis of assumptions but what we see in space is completely out of date. Scientific researchers, when describing the Cosmos, often speak in the present tense which is a bit unreasonable since the events they use as references took place billions of years ago. I quote Dave Rothstein, a former graduate student and postdoctoral researcher at Cornell who used infrared and X-ray observations and theoretical computer models to study accreting black holes in our Galaxy.

'When light is emitted from one galaxy and travels through space to another galaxy, during its trip through space it also will be stretched, causing it to have a longer wavelength and therefore causing its colour to appear more towards the red end of the spectrum. This is what leads us to see redshifted light when we look at faraway galaxies, and it is measurements of this redshift that allow us to estimate the distances to these galaxies.'

Scientists conject that the reason why we can identify celestial bodies at much further distances than 13.8 billion light years, the presumed age of the Universe, is because it is expanding. They conclude that matter/energy accelerated away from the 'Big Bang' and is still expanding. This 'centrist' theory creates an enigma since it causes other problems to raise their ugly heads. The thing about us human beings is that we don't like to be proved wrong. Consequently, we dream up more and more improbable reasons why we are right rather than opening our minds to alternatives.

In my opinion, the Cosmos has neither a 'centre' nor dimensions. Frank Heile, Ph.D., disagrees: *'If we waited for 46.5 billion years, we should actually be able to see the light emitted right now from those superclusters of galaxies in our telescopes. The light has just now started on its way towards us but it will take a while to reach us since it will have to come from the surface of a sphere with a diameter of 93 billion light years - with us at the centre! Unfortunately, it is no longer true that we shall eventually see that light from those super clusters.*

'The problem is that we now assume that due to dark energy, the expansion of the universe is actually increasing at an accelerated rate (actually it has been said to be accelerating for at least 3 billion years). Because of the accelerated expansion, those superclusters, which are now 46.5 billion light years from us, will be receding from us at a rate that is greater than the speed of light by the time we wait another 32.7 billion more years. So, the diameter of 93 billion light years is, at most, a theoretical estimate...'

One point I must make here is that if a radiant object is departing from an observer at greater than the speed of light, Relativity will mean its light cannot be received by him! For example, if a target recedes from a marksman's rifle at greater than the speed of a bullet, the target will be unharmed. The bullet will drop harmlessy to the ground when its energy is reached. I question the assumption that we can calculate the dimensions of the Universe for the foregoing reasons but in my opinion the subject is irrelevant. The Cosmos is better described as an infinite 'closed system' of energy/matter. Quantum Theory suggests that theoretically, space doesn't exist. Time and space may just be aspects of organic entities which have developed using only electromagnetic energy thus limiting their communication and development to the 'pedestrian' speed of light. Ironically, we share our existence with particles that are truly 'free'. Einstein was conceivably the greatest scientist of all time. He maintained that light cannot travel faster than 186,000 miles per second and he was absolutely right as far as the visible thermodynamic universe is concerned and consequently he formed the theoretical equation $E = mc^2$ where c is the constant being the speed of light and electromagnetic forces. But, this raises questions about how the inflation of the Universe following the 'Big Bang' was supposedly at greater than the speed of light and is still accelerating. How else can we explain why there are celestial objects up to 45 billion light years away, older than the presumed age of the Universe, which is 13.8 billion years? I think I can!

The term 'inflation' has been further interpreted as space being 'blown up' like a balloon when objects on its surface will separate. You can test this by putting a couple of dots with a biro close together on a new balloon and then blowing it up. The dots will separate. In relation to the 'Big Bang', this raises the problem of where the 'breath' or energy came from to inflate the Universe? Mathematical calculations leave equations an enormous amount of energy short. The next step was to call it 'dark energy' which is an adequate but unintentional way to describe Quantum energy.

There is a distinct difference between electro-magnetic energy and 'free' Quantum energy. Light is composed of photons which are its basic unit of electro-magnetic force and its speed cannot exceed the above figure. I suggest that 'free' or 'dark' energy is comprised of sub-atomic particles such as electrons with no discernible mass (some people describe them as only electrical 'charge'). However, they can also exist as integral parts of larger material bodies such as atoms orbiting their nuclei under conditions governed by Thermodynamic Law. To do so they must forfeit their Quantum nature.

The only difference between the equation $E = mv^2$ which is used to determine the kinetic energy of a moving body and Einstein's $E = mc^2$ is that v is a variant and c is a fixed constant. Whichever equation is used, energy will always be the reciprocal of mass (matter) and vice versa whatever value is ascribed to either c or v. On that basis, v becomes the most important factor and provides the means to look at the Cosmos through 'new eyes'. Look at what happens when mass becomes infinite (∞) and energy = zero (0): $v = 2\sqrt{(E/m)} = 2\sqrt{(0/\infty)}$ = zero. Now see what happens when energy becomes infinite and mass zero: $v = 2\sqrt{(E/m)} = 2\sqrt{(\infty/0)}$ = infinity.

Something quite bizarre is revealed. When 'free' energy becomes infinite, so does its velocity and when 'free' energy becomes zero, velocity does too. Apparently, e is equivalent to v under all circumstances! Can we assume therefore that infinite velocity is the 'natural state' of 'free' Quantum energy, when it is independent of matter? What this suggests is that zero matter (or particles of matter so small that they behave as if they have no mass) travelling individually at an infinite speed can equate to an infinite amount of energy.

If the energy of a stationary but infinite amount of mass is zero, it follows that the energy of zero mass moving at an infinite velocity must be infinite. The use of the word 'relative' when describing infinites may be advisable since only the Cosmos as a whole is truly infinite. Can we look at the velocity of material objects as the sole determinant of their 'quality' and thus their 'place' in the entirety of the Cosmos?

Since this point is very important, I will equivocate: In a Quantum state, effectively the mass of subatomic particles becomes equal to zero and their velocity becomes infinite. Movement = energy. There is no energy in a mass if, including its component parts and the orbiting electrons within its atoms, is totally stationary. Only the centre of a 'black hole' can provide the force necessary to bring everything to a halt in a thermodynamic environment when matter can instantly be converted to energy or purely movement. There is a huge amount of 'free' energy in zero mass (or a mass broken up into such infinitesimally small particles that individually they have no mass), moving at an infinite velocity).

Infinitesimally small particles with no mass are unaffected by gravity until they eventually sacrifice their immunity in the creation of new matter. They can either combine with one another to create matter or unite with existing matter. 'Free' can equally mean 'unattached'. An eternal 'Cosmic see-saw' governed by the Law of Conservation of Energy continuously works to maintain Cosmic Equilibrium.

It's reasonable to assume therefore that a 'state' of Singularity is where 'dark' or 'free' energy resides. This opens the possibility that Singularity is comprised of 'free' energy without detectable mass. One might argue that there are intermediate conditions when 'free' energy's velocity is faster than that of light but less than infinite but my thinking is that to escape from black holes, energy must become Quantum or Singular.

The terms are not differentiable. We know that electrons are affected merely be being observed. This is a reasonable assumption since they will have been changed by an encounter with an observer; just as an observer will be changed by encountering energy! "Schrodinger's cat"?

Generalised 'Centrist' theories of the Cosmos are remindful of when we thought of the Earth as being the centre of the Solar System around which everything else circled. It looks as though we still do! This doesn't surprise me when our egos continuously encourage us to believe we 'are the centre of the Universe', a metaphor for celebrity status sometimes used as a derogatory description of someone we dislike.

'Centrism' leads us to believe that there was a 'Big Bang' event when the contraction of all matter into an infinitesimal spot created a huge explosion. We then assumed that 'inflation' expanded energy into the universe around us and the matter which it gave rise to has continued to accelerate away from us at greater than the speed of light. This is meant to explain why there are celestial bodies older than our Universe.

There are much simpler explanations staring us in the face which support Einstein's $E = mv2$, but not by imposing the constant 'c'. Our view of the Cosmos/Universe is far from its reality. We co-exist with a Quantum Singularity in which space and time are indefinable and 'free' subatomic particles occupy it as dark energy. When these either interact with one another or with existing matter, they become at least temporarily bound with, and subject to the Thermodynamic Laws which govern the material world.

Currently, it seems difficult to mathematically compromise Quantum Theory with Thermodynamic Laws. This might become possible if we accept that 'free' energy can be created from matter if it is subjected to sufficient gravitational force in a 'black hole(s)'. This would de-construct it and 'squirt' free energy back out at infinite speed. Bearing in mind that infinitesimal particles exist in a state of Singularity obeying Quantum Theory would mean they are all effectively everywhere at once or conversely, 'in the same place'.

I'm suggesting that Space and Time in a Quantum/Singular state don't exist; they are phenomena peculiar only to the 'phony' reality of a world of matter and governed by Thermodynamic Laws. 'Black holes' are 'recycling points of transformation'. Take a look at the following which was posted on the Cern website: 'The Large Hadron Collider (LHC) is the world's largest and most powerful particle accelerator. It first started up on 10 September 2008, and remains the latest addition to CERN's accelerator complex. The LHC consists of a 27-kilometre ring of superconducting magnets with a number of accelerating structures to boost the energy of the particles along the way.' This Boson particle was apparently identified as a result but is still a hypothetical particle.

By artificially applying a colossal amount of electromagnetic energy to a relatively 'closed' system, Cern might just have created some matter in the process. Perhaps they should be celebrating how they have managed to produce a new particle instead of claiming they have discovered an existing one. So far, proof of the existence of the Boson is mainly theoretical. However, the research undergone there adds some strength to my own admittedly controversial theories. Wikipedia says this: 'The positron or ant electron is the antiparticle or antimatter counterpart of the electron ....and has the same mass. When a positron collides with an electron, annihilation occurs......certain kinds of particle accelerators involve colliding positrons and electrons at relativistic speeds. Their high impact energy and 'their' (sic) mutual annihilation..... creates a fountain of diverse subatomic particles.'

Surely, what is described above is replicated within 'black holes' where an infinite gravitational force crushes orbiting electrons and positrons together to cause their annihilation into countless subatomic particles effectively without mass, from which the resulting energy can only be expressed as velocity. In the 'immediate surroundings' of 'black holes', the density of these particles will be so great that many will unite immediately making them constituents of matter/energy visible to us as 'geysers' of emissions. However, the overwhelming majority depart as 'free' energy into the state of Singularity/Quantum Theory where infinite opportunities to 'rematerialise' await them.

' Since publishing my book entitled "Einstein's E = mc^2 unravelled", I have discovered that serious challenges have been made to the method of measuring 'red shift' to determine galactic distances. This is the principle reason which allowed Georges Lemaître and subsequently the 'Big Bang' theory to gain credibility. Scientists' 'cast iron alibi' is that they can supposedly calculate stellar distances by using 'red shift'. Put simply, the red part of the light spectrum is known to expand with distance so by measuring that, it's assumed we can calculate a star or galaxy's distance from Earth.

However, 5 years ago, research performed by a team of astrophysicists led by Eric Lerner from Lawrenceville Plasma Physics, resulted in a headline 'Universe is Not Expanding' in the magazine Science News. It emphatically demonstrated that whilst the theory that the 'red shift' method can be used to determine the distance of relatively local stars and galaxies, it is unreliable at much vaster distances.

The International Journal of Modern Physics examined their work and later commented: *'We do not claim that the consistency of the adopted model .....is sufficient by itself to confirm what would be a radical transformation in our understanding of the cosmos. However, we believe this result is more than sufficient reason to examine this combination of hypotheses further'.*

There is no such thing as Time

For some inexplicable reason, scientists have convinced themselves that Time is tangible. In doing so, they have created a series of dilemmas because once one's mind becomes saturated with a notion, it's almost impossible to let go of it.

If something exists at all, by reasonable deduction one has to be able to describe it in rational terms. Millions of philosophical hours have been applied to deciding whether objects actually exist or are merely fragments of our consciousness. We can assume without much doubt that a chair exists because it conforms to so many commonly substantiated expectations. It has an agreed shape, is firm to the touch, can support weight and most of its many forms are logically acceptable.

We've tried very hard to describe Time in rational terms. The truth is that it only possesses the value an individual ascribes to it. Actually, Time is immeasurable since it is entirely subjective. Its duration is appropriate to how a passage of events is experienced by any living organism. The nearest we humans get to it is by counting our heartbeats but even they beat at regularities which vary according to our experience of being conscious. If we are asleep, time doesn't exist.

The purpose of having a theoretical yardstick of Time is merely to create standard durations of it. These apportion intervals with which to label events. Unless events have Time labels, they are just a meaningless jungle. Homo sapiens and most living creatures have structured their lives around natural events such as days, lunar months, years and seasons. Early philosophers ascribed 24 hours as the period of time separating one day from the next. Nowadays, intervals of Time are measured by atomic clocks. We use the frequency of electronic transitions in certain atoms to measure the second. The International Standard of Time defines the second as 9,192,631,770 cycles of the radiation that corresponds to the transition between two electron spin energy levels of the ground state of the ^{133}Cs atom. A second is only 1/3600 of an hour so it's still linked to what is a local event, the Earth circling the sun every 365 and a ¼ days and spinning on its axis every day.

This crude method of providing intervals to separate events is only relative to the medium in which living organisms experience their lives. It's hard enough for humans to relate to events occurring in the Universe when the medium by which we experience them is light which travels at only 186,000 miles per second. A bat is far worse off in relation to reality; it lives in the medium of sound so will always be 'in an even slower lane'. Without the means of accurately coordinating everything into one 'reality', we organic creatures are quite incapable of sensing the Universe as it is. Our experiences of everything 'out there' is like looking into history. Why wouldn't our physicists have declared that nothing can exceed the speed of light when that represents our only view, sense and conscious experience of the entire source of our being; light?

We didn't decide to create a way to measure time because it's a tangible substance. We chose to link our solar system's functions to events and that is still our only way of comparing our locality with any other in the Cosmos. This hasn't stopped us however, from making endless assumptions not only of where we are in relation to everything else in the Universe but also what is happening wherever they are! The fact is that we don't know if they even exist.

We have attempted, vainly in my opinion, to calculate how far distant objects are without knowing how the light that precedes them has reached us and equally importantly, from what direction it originally came from. Gravity bends light. To cap it all, we have interpreted space as constituting a vacuum containing only time; the greatest self-delusion we could invent. We have mapped space using only an interval of time that we invented only to keep our memories in chronological order.

T me only exists as a storage bank of individual events tied to the dates and times of their occurrence which we have ascribed to them. This is only moderately useful when events occur at 'arm's length'. When they take place at great distances, all we view is their history.

Data banks in computers are useless unless every piece of information is linked to a specific date and time. Our brains are full of memories but these are useless to us unless we know when the events we have recorded occurred. What is the point of saving a piece of nformation if it isn't linked to a specific time? Without that link, it would be only free-floating data with no means of understanding its purpose or role in a story. A CD disc can contain a huge amount of data. If we play some music or watch a movie without that data being supplied to us in a regulated manner as closely balanced to the events in the order in which they occurred, it would become an unintelligible cacophony of sound and images. It's our brains which limit our knowledge of the Universe since they like us are bound to the speed of light. Or, are they? That's the question!

I₋ we were incarcerated in an isolated environment, we would have no idea how quickly or slowly Time was passing. A sleepless night or a boring day can 'last an eternity' but a trip to the seaside with a companion can be over in a trice. Ironically, we need to 'clock-watch' in order to judge our happiness! Are we happily engaged or slowly hating what is going on! Unless we learn to meditate and slow ourselves down but make time go faster, having nothing to occupy us can take an eternity for Time to pass. Every life-form has its own life/time story. Each one records a unique history of its life with labels linking everything that happened in it to an entirely artificial <u>rate of change.</u>

Einstein became preoccupied with Time to the extent that he believed space itself was bent by gravity. I suggest that it is not galactic entities themselves that are interfered with by Time, it is only their images. When we look at the sky, we cannot see the reality of any actual stellar objects; we see only their distorted images.

It amuses me that science fiction often includes episodes when its 'superhuman' characters either 'go backwards or forward in Time'. Neither one of those options is open to us. Please don't listen to abstract fictional theories describing these actions as being credible. We are in one part of the Universe and every other entity is elsewhere. Even the sun goes down 8 minutes before it gets dark here. We organic mortals cannot even see where they were when their light left them or how far away they are; or whether they still exist. Similarly, observers on them cannot see the Earth as it actually for the same reasons.

An amusing thing that scientists have mistakenly deduced is that if a person travels at a great velocity, time for him will 'stand still'. Evidence for this appears to include the fact that an airline pilots' watch and even a computer will slow down at aeronautical speeds and the amount is measurable. This is hardly surprising since any watch or measuring device situated in a changing environment will experience changes in the physical conditions in which it functions. These will vary between high rates of acceleration and deceleration all of which will cause gravitational effects on its mechanism. To assume then that the pilot's life will be extended even marginally is ridiculous. All that has happens is that the measuring device and his body will have slowed down.

We use the term, 'x light years away, as a means of using Time to provide us with some knowledge of an object's distance, even dare I say, its 'state'. Equally foolishly, we use distance travelled by light from distant bodies (measured incidentally by means of light's red shift or its luminescence both of which are unreliable measurements at great distances) and assume it is the same as the object's radial distance. You can read some of my other submissions concerning the Big Bang to see what I think of that idea!

Let us imagine that someone somewhere has invented a means of travelling at an infinite speed and you can use it to see what is really out there. You aim your 'time machine' at an interesting place full of stellar objects about 14 billion light-years from Earth and press the switch. Surprise, surprise, you find nothing there! Back to Earth in a second, rewind and retarget an area about the same distance from earth but completely empty. This time you arrive at a place full of galaxies, planets and stars!

What you are doing in each case is certainly not visiting the future. You are looking at what already existed but on Earth you would have had to wait 14 billion years to experience! Bear in mind however, that from wherever you view the Universe, all you will get are out-of-date versions. The farther away objects are the more 'behind' them you will be. So, this means that wherever we go we will be experiencing reality only where we are.

Reality does not exist if there is any distance from an observer to any object he views. A bat only experiences the reality of a moth when it's in his mouth. Until then it's just a vibration in its 'eardrums' which falsely indicates its whereabouts.

Compared to your pre-launch 14 billion years out-of-date view from Earth, if you travelled there in a fraction of a second you will catch up only with its actuality. Look back at Earth and you will 0nly see how it was 14 million years ago. Wait where you are for another 14 billion years and you will only see it as it was when you left it. You'll still be out of touch with it by a long way.

On the trip I fantasised, you haven't stepped into the future at all; you have stepped into the reality of a different place. Look at it the other way around. If you decded to stay there, all you will see looking back will be Earth's past. Of course, if you wanted to see it as it is right now, you would have to defy your mortality and wait about 14 billion years until its 'past' (including you setting off) will arrive at your new abode but that will by then be 14 billion years out of date! You might see yourself leaving!

Going back in time and changing the future is nonsense. Given the possibility of travelling at an infinite speed, one can still only view where you are with any degree of reality. The point I am making is that the only reality we experience is when we are in very close proximity to objects, articles, organisms, which we observe. If we observe them from far away, we look at their histories.

We organic life-forms will never know what constitutes reality since for us it doesn't exist. Instead of realising this fact, scientists have spent trillions of hours trying to examine the Cosmos and still talk of it in the present sense.

Getting to the point of this 'mind experiment', everything we experience is related to how our sense of it is communicated to us. Obviously, we cannot be in every place at once but that's only because we have a very poor system of communication. We can't even play chess with someone on Jupiter. It would take 86 minutes to learn how he had reacted to your "Check!"

move. A game could last for 7 days not counting thinking time! You could be playing with someone who had died. A God would never be able to work under those circumstances. He needs to know what is <u>currently</u> going on in every corner of the Universe.

This brings me to my concluding suggestion. Even if the Universe is not infinite in size (I believe it is), the distances we know exist, make communication with its extremities impossible to accomplish using only light and electromagnetism as our means/medium of communication. If, matter can only communicate with everything else at the pedestrian speed of light (as seems the case currently), the 'time discrepancies' between them all would be enormous. But that doesn't seem to be the case. Every direction we look in shows a uniformity of development 'order' and 'chaos'.

In my book, "Einstein's $E = mc^2$ unravelled – an alternative view of the Cosmos", I make the case that Thermodynamic Laws apply even when matter becomes capable of speeds vastly exceeding that of light. I postulate that black holes are merely 'recycling centres' which balance entropy and extropy even within galaxies. Our own galaxy contains hundreds if not thousands of them. Without their actions in turning matter into energy, there would be no Universe. It would have contracted into one inescapable point.

Philosophers such as Leibnitz and Spinoza have suggested that there are similarities between Nature and God in the way the Universe functions. However, it must surely be obvious to my reader that for man to view the Universe as it is with apparently very little differences in its make-up over what are to all intents and purposes, infinite distances, we must possess the ability to view things 'as they are' and not 'as they were'.

Also in my book, I talk about the ability of Quantum particles such as electrons to communicate instantly with one another over infinite distances. We play at understanding their language; it's not within our ability to do so. Nevertheless, hints are provided to us from time to time by means of Synchronicity, the term ascribed to inexplicable coincidences by Carl Jung. One thing is certain; we share electrons with everything there is. We may have much more to learn about them before we start setting off to explore a Cosmos that doesn't really exist as anything more than an enticing illus on.

A final thought! If it takes 8 minutes for light from the sun to reach Earth where did that 8 minutes go. We couldn't see it leave until 8 minutes after the event and a sun dweller couldn't see it arrive here until 8 minutes after it did so. Even travelling at the speed of light wouldn't help us. Only the simultaneous synchronous experience of everything sometimes termed Singularity makes it possible. Singularity is neither a point in which everything exists nor is it an all-encompassing space.

Neither exists. Communication in Singularity would be instant. Using only the medium of light is just a distraction.

Do we intend to offer a logical fight to Tackle Climate Change or surrender to the consequences of our own stupidity?

This is not intended to be a political observation and should not be assumed to be so. It is a factual statement of how the world has acted in the past and continues to do so now. It is blatantly clear to me that the forces of greed which have brought us to the dark future that awaits us will not stop their fight to make a profit out of disaster. Just as the tobacco barons who still fight rationality and common sense, so will the petrochemical industry continue to find ways its carbon-based fuels can continue to poison our lives. They had their day but like coal must now retire gracefully.

We cannot trust politicians to act sensibly. Most of them have few practical abilities and usually 'follow the money'. In the UK, our government put our lives in the hands of some of their 'mates' to handle critical Covid-19 activities such as the provision of Personal Equipment and the management of Track and Trace with abysmal failings.

Recently, our Prime Minister began to champion the use of Heat Pumps as way to reduce energy consumption. For every 3 kWh of heat they provide, 1 kWh has to be imported. A typical home requires about 25 kWh to heat it. Thus 8+ kWh have to be supplied from the national grid. At a voltage of 250 AC that means a current flow of about 35 amps. It doesn't take an Einstein to work out what effect a few mill on homes pulling that sort of power would have. The grid infrastructure would be knocked out in just the same way that electric cars would decimate it.

Please read what I have to say below and join me in uniting the world in a crusade of international action. We can restore its fading beauty in a truly shared programme of logical, rational action. It just takes some common sense! Let's stop tinkering with the challenge and get down to the meat of it. We have more than enough of the only ingredients we will need; sunlight and water!

1. I have attempted in this paper to offer a practical solution to the problem of Climate Change. Clearly, only global action will ameliorate the dire situation the world is facing but we must strive as a nation to do everything in our power as quickly as possible to reduce our own carbon footprint. I believe no other proposals I have studied can accomplish this either as quickly or

as financially attractively as the one I suggest. Of course, other countries will have varying economic structures but basically it is a 'fit-all' global solution.

2. I have provided the evidence to substantiate why the most expedient method available to the UK is by using the sun's energy to accomplish this by the use of domestic and commercial solar panels. One hour's duration of the energy which reaches Earth from the sun each day would power the entire planet's needs for a whole year!

3. The main problem associated with using green energy from both wind and solar sources is how to store it. This can be overcome as I describe below by using this 'free' electrical energy to manufacture hydrogen by the electrolysis of water. This can be liquefied and stored for use at our discretion. I have fully described how this process works and how the associated infrastructure costs are financially feasible.

4. Ironically, the most important of my findings is that by following the route I prescribe, an enormous financial benefit will accrue to the UK's economy which makes the programme worth carrying out on that score alone.

5. An initial expenditure of £30 billion spent over 5 years and financed through existing low interest rates would allow us to install the first 10 million domestic solar panel installations. An estimated 300,000 mainly outdoor jobs would be created almost immediately which would boost our economy and be largely unaffected by Coronavirus restrictions. Remember, we have spent many times that amount already to fight Covid-19!

6. This first 5 years expenditure would be more than matched by the concomitant savings made over the next 25 years, the expected life-span predicted for solar systems. In that period of time the power thus generated will save the country an estimated £112 billion based on current electricity costs. Multiples of such programmes could be repeated to match sustainable energy requirements.

7. Just the action I prescribe would reduce our CO2 emissions over the next 25 years by a massive 1.75 billion tonnes.

8. We must not overlook the associated benefits of better public health and the ability of people to be hopeful for their future at one of the direst times in our history. I presented my ideas extremely briefly at a local Climate Change

meeting a year ago. Several people approached me afterwards saying I had given them hope in the realisation that there is actually something tangible that can be done.

9. In an age when many jobs are being replaced by robots, this proposal offers significant work opportunities which will remove hundreds of thousands of people from the unemployed registers.

Please take a look at what the environment correspondent of the BBC said recently on the link https://www.bbc.co.uk/news/science-environment-48964736
I quote from it:

> *Do you remember the good old days when we had '12 years to save the planet'?*
>
> *Now it seems, there's a growing consensus that the next 18 months will be critical in dealing with the global heating crisis, among other environmental challenges.*
>
> *Last year, the Intergovernmental Panel on Climate Change (IPCC) reported that to keep the rise in <u>global temperatures below 1.5C this century</u>, emissions of carbon dioxide would have to be cut by 45% by 2030.*

But today, observers recognise that the decisive, political steps to enable the cuts in carbon to take place will have to happen before the end of next year.

The idea that 2020 is a firm deadline was eloquently addressed by one of the world's top climate scientists, speaking back in 2017.

Why is it so hot and is climate change to blame?

"The climate math is brutally clear: While the world can't be healed within the next few years, it may be fatally wounded by negligence until 2020," said Hans Joachim Schellnhuber, founder and now director emeritus of the Potsdam Climate Institute.

The sense that the end of next year is the last chance saloon for climate change is becoming clearer all the time.

"I am firmly of the view that the next 18 months will decide our ability to keep climate change to survivable levels and to restore nature to the equilibrium we need for our survival," said Prince Charles, speaking at a reception for Commonwealth foreign ministers recently.

The Prince was looking ahead to a series of critical UN meetings that are due to take place between now and the end of 2020.Ever since a global climate agreement was signed in Paris in

December 2015, negotiators have been consumed with arguing about the rulebook for the pact. But under the terms of the deal, countries have also promised to improve their carbon-cutting plans by the end of next year.

Residential and industrial heating processes contribute a huge amount of CO_2 (carbon dioxide) to the atmosphere. I did some calculations last year in connection with CO_2 emissions arising from transport and established that an average car contributes about the same amount of CO_2 as heating and lighting an average home. This surprised me but the pie graph below confirms my findings:

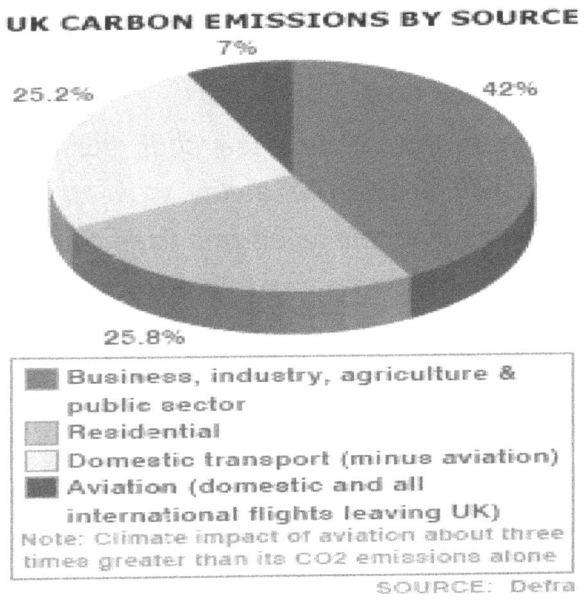

UK CARBON EMISSIONS BY SOURCE

7%

25.2%

42%

25.8%

Business, industry, agriculture & public sector
Residential
Domestic transport (minus aviation)
Aviation (domestic and all international flights leaving UK)
Note: Climate impact of aviation about three times greater than its CO2 emissions alone

SOURCE: Defra

This strikingly reveals that domestic heating and lighting together with domestic transportation are responsible for over 50% of the country's CO_2 emissions! I've looked at many other graphs showing the numerous ways in which we pollute the air with CO_2. I believe these encourage us to seek 'piecemeal' solutions for various sources of emissions rather than seeking a broader and more effective overall solution as I suggest.

Logic dictates that if we are to urgently tackle climate change, we need to find a plan of action that has an important element common to all of our problems and which links them together into one major solution. The production of hydrogen by the hydrolysis of water using renewable sources of energy offers such a means and is the reason why it has interested me for about 20 years.

The link https://www.carboncommentary.com/blog/2017/7/5/hydrogen-made-by-the-electrolysis-of-water-is-now-cost-competitive-and-gives-us-another-building-block-for-the-low-carbon-economy describes how efficient this process is.

I quote from it:

"Hydrogen from electrolysis

The world produces about 50 million tonnes a year of hydrogen. (Some sources suggest it is more than this). The gas is used as an additive in oil refineries, as a raw material for making ammonia and for many different industrial processes including, for example, the making of margarine.

Almost all hydrogen is made today from what is known as 'steam reforming', usually of methane (the main constituent of natural gas). A stream of gas is mixed with high temperature steam in the presence of a catalyst. The eventual output of the process is a mixture of CO_2 and hydrogen. The valuable hydrogen is collected and the CO_2 vented to the atmosphere. If my calculations are correct, the hydrogen produced today through the steam reforming process is resulting in approximately 500 million tonnes of emissions a year, or well over 1% of global GHGs.

Hydrogen can also be made using electrolysis of water. Electricity is used to split the molecule into hydrogen and oxygen. If made using water electrolysis, global hydrogen production would today use about 15% of world electricity generation. When manufacture of H2 is switched from using methane to employing surplus electricity, hydrogen will be an important method of balancing the world's grids. When power is abundant, the electrolysers will be turned on. Their work will stop when electricity gets scarce.

In the past, electrolysis was very rarely employed because the energy source, electricity, was more expensive than the gas used for steam reforming.

Is this still true? We need to investigate the energy efficiency of steam reforming and its operating and capital costs as well as the relative prices of gas and electricity.

Very roughly, a new electrolysis plant today delivers energy efficiency of around 80%. That is, the energy value of the hydrogen produced is about 80% of the electricity used to split the water molecule. Steam reforming is around 65% efficient."

We don't have the time to explore the myriads of innovations currently on offer when the hydrolysis of water to produce hydrogen has been an established solution for a considerable time. Although many entrepreneurs are seeking ways to promote their own methods and/or their equipment, I would urge the government to concentrate its time on seeking ways in which this proven method of reducing CO_2 emissions can be exploited in an urgent programme of change.

There are three main sources of CO_2 emissions:
1. The business, industry, agriculture and public sector which includes a substantial element of travel

2. Residential

3. Domestic travel

Over the last couple of decades, the sun's energy has been increasingly utilised by converting it into electricity using solar voltaic panels. These have become progressively cheaper through mass production favourably changing their economic perspective.

This has made solar voltaic panels not only the principle means of reducing CO_2 but equally importantly emphasised their economic advantages

The Domestic Energy Consumer
From https://www.cat.org.uk/info-resources/free-information-service/energy/solar-photovoltaic an average house with a suitable aspect can now be equipped with an array of solar panels for as little as £5,500 capable of generating a peak output of 3.75 kilowatts, amounting to approximately 3,000 kWh per annum. I have seen 4 kW installations offered at £4,250

but prefer making a conservative case for my arguments.

An average 3.75 kW installation is currently guaranteed to work reliably for at least 20 years only losing during that time up to 20% of its generating capability. With bulk buying on a much larger scale than presently, I surmise that the cost of an average installation could be reduced to about £3,000. This is equivalent to £300 p.a. over a normal 10 year write-off period. Currently, a person who has installed solar panels on his roof exports about 50% of the electricity generated for which his energy supplier pays him about 3 pence per unit which amounts to £45 p.a. Additionally, he will use about 50% of power generated, 1,500 kWh p.a., to replace what he would have spent on buying it from his energy supplier at about 16 pence per kWh. This has saved him a further £240 p.a. After ten and a half years, he has broken even and thereafter saves about £285 p.a. for the life of his installation also rendering him largely immune from energy price inflation.

For every domestic customer who installs a solar panel array, another's needs are satisfied because each domestic installation exports about half as much as its owner consumes. Thus a saving in CO_2 emissions of 7 tonnes p.a. occurs with every domestic solar panel installation.

Internet sources advise me that so far only 800,000 houses have installed solar voltaic panels. Effectively at present therefore, they have reduced our CO_2 emissions by a staggering 5,600,000 tonnes per annum.

The following link suggests that we should be aiming at the equivalent of 10 million houses with solar voltaic installations.

https://energysavingtrust.org.uk/blog/present-and-future-uk-solar-power?gclid=Cj0KCQjwsuP5BRCoARIsAPtX_wFiT6lwHA6ODKo5T26N-IBFZ4Y43f4W1da26WHK5Y6rgi_pWK31JpYaAi7xEALw_wcb

Most houses use natural gas to heat their homes. In this respect, replacing it with hydrogen can eliminate the emissions of CO_2 from their boilers. Hydrogen is already being introduced into gas distribution systems. It has been produced either by the electrolysis of water described above or by combining methane (CH_4) with water (H_2O) to form a mixture of CO_2 and Hydrogen from which the later is extracted. This process is less expensive to construct but less effective.

The following link deals with the technicalities of using domestic boilers when fuelled by hydrogen. It is encouraging to say the least.

https://www.boilerguide.co.uk/articles/hydrogen-boilers-alternative-gas-central-heating

The Energy Supplier

According to the government website https://www.ofgem.gov.uk/data-portal/wholesale-market-indicators the average price paid by our energy suppliers on the wholesale market is about £50 per megawatt hour for electricity or 5 pence/kWh. An enormous opportunity exists for them to obtain their supplies at a reduced and reliable price by engaging themselves in promoting and even financing domestic solar panel installations. At present, they import power from domestic sources at only 3 pence a kWh. By entering into contracts with domestic and commercial property owners, I'm sure it would pay them to loan their customers the cost of installing solar panels in return for abandoning paying them for any exported electricity. Their main source of income would remain with their charge for electricity consumed by their customers.

What's more, financial inducements by the government would become unnecessary. The process would become self-financing.

The Domestic Car Owner

I fear that too much store has already been placed in using electric cars to solve our problems of CO_2 emissions without fully understanding their shortcomings.

- They are still too expensive for the pockets of the average car owner

- At the time of writing, there are still only about 40,000 fully electric vehicles in use out of some 33,000,000 privately owned vehicles

- The infrastructure is completely unable to deal with the millions of battery-fuelled vehicles which would be needed to make a real difference. Most car owners do not have off-road battery charging facilities. Allowing hundreds of thousands of electric cables to stretch across pedestrian pavements is out of the question.

- The only way that electric car charging can be facilitated in numbers great enough to make them truly feasible would be to cease trying to manufacture batteries for them with huge distance capabilities. Instead, make electric car manufacturers globally standardise on one type of rechargeable battery which is interchangeable between all electric cars. Filling stations could then move from being fossil fuel distributors to battery replacement centres with their own daytime re-charging facilities powered by renewable energy.

- We must urgently encourage workers and their employers to adapt permanently to far less centralisation. Covid 19 has shown how millions of us can comfortably work from home. Perhaps we should be putting job performance before workers' timed physical presence?

- Car-sharing must be encouraged and more 'Park and Rides' made available.

- We must exponentially increase the number and use of electrically-powered buses, coaches, cycles, motorcycles and trains.

- The use of private cars in town and city centres should be progressively discouraged forcing people to acclimatise to public transport or use Park and Rides thus stimulating investment in public transport opportunities. In my own town, a once active P&R is now disused mainly due to inadequate bus services and Council policies which do not encourage them.

- It would be a good idea to reduce the national maximum speed limit to 50 mph nationally. This was done in 1976 to overcome fuel shortages. It would reduce vehicle fuel consumption by an estimated 15-20% reducing not only our CO_2

emissions but also our oil imports. Incidentally, this has been shown to also reduce overall journey times.

The Government and the Economy

The following link confirms my views concerning the practicalities of using hydrogen to replace natural gas. https://www.carboncommentary.com/blog/2017/7/5/hydrogen-made-by-the-electrolysis-of-water-is-now-cost-competitive-and-gives-us-another-building-block-for-the-low-carbon-economy

From information provided in https://notalotofpeopleknowthat.wordpress.com/2018/03/16/uk-natural-gas-imports/ I am able to make some useful deductions.

In 2017, we had a net import of Natural Gas totalling 34.3 million tonnes of TOE (tonnes of oil equivalent. From an internet source, "*Gas is one of the key pillars of the UK's energy mix, accounting for 29 per cent of the UK's energy production and second only to oil. Gas production from the UK's Continental Shelf (UKCS) would have been sufficient to meet nearly 50 per cent of UK demand in 2019. Gas is particularly important for electricity generation where it meets around 40 per cent of the fuel required in power stations. It is also critical for space heating, domestically and in offices, hotels and restaurants. In 2019 gas met nearly two thirds of total domestic energy demand.......*"

The cost of natural gas fluctuates significantly on the international market and is currently about £2.24 per MMBtu (million British Thermal Units). This means our net imports of gas are costing the economy about £2.4 billion per annum. If our net importation of natural gas was entirely replaced by the production of hydrogen using solar power, the cost of converting electricity to hydrogen would be largely met by its savings. Amazingly, natural gas costs only about 0.8 pence per unit (kwh) to import but energy suppliers make a huge profit since most domestic consumers charge about 7 times that figure.

I was urging a decade or so ago that energy tariffs should be progressive. Currently, large consumers pay less per unit than the smallest. This is irrational. I would suggest that in order to reduce CO_2 emissions, we must encourage frugality and provide the incentives to do so. Only by linking fuel extravagance to higher tariff prices will this be accomplished.

The most important aspect of my proposal is the sheer economic advantage arising from the measures I have suggested.

The main problem with renewable energy and its solution

The generation of electrical power from solar panels is relatively easy. The real problem has always been how to store it. The principle ways of generating renewable energy by wind power and solar panels are susceptible to weather conditions. Solar panels can only generate power during about 8-10 hours of the day with varying production rates depending on their geographic location and aspect. It is imperative that we quickly establish and put into place a means to store the electrical power they generate.

Generating more solar power than we need to run our economy during sunlight hours will allow us to use the surplus to hydrolyse water into hydrogen and oxygen, its two constituents. The hydrogen can be licuefied and stored as a 24 hours a day standby. This would balance supply and demand by cleanly fuelling electricity generation at power stations using gas turbines connected to the national grid.

More information on solar panels is available in the link https://www.cat.org.uk/info-resources/free-information-service/energy/solar-photovoltaic/

Some very important points made by my son suggest that the "UK's renewable energy profile is somewhat imbalanced at the moment with world-beating offshore wind but unexploited rooftop solar power opportunities. More solar power might, for example, have obviated the need to use a coal-fired power station during the August 2020 heatwave. More generally, rooftop solar power would be produced at the same time as maximum future demand on the electricity network for air-conditioning, which more people will be using in future".

In this regard, perhaps it should become mandatory for all new houses and commercial premises to be equipped with solar panels provided the sites are technically suitable. In the case of the average house, the cost of their installation would represent a mere 2% of its market price and could justifiably be covered by a government grant.

Last but not least, already a significant number of trains and large road vehicles used for public transport and the carriage of goods and materials are now being equipped with hydrogen fuel cells and electric motors. Without the space restrictions of smaller vehicles, hydrogen can be carried on board in pressurised tanks to fuel the cells which convert hydrogen into electricity on demand. The use of a gaseous fuel such as hydrogen to fuel cars is not novel. During the Second World War, fuel shortages encouraged the use of coal gas to drive cars converted from petrol. The coal gas was carried in huge bags attached to the roofs of the cars. Need is the mother of invention! Research is already underway to develop hydrogen tanks small enough for cars. It's only a matter of time before filling stations will provide not only battery charging and replacement facilities but hydrogen 'pumps'. Quite rightly, fossil fuels will have had their day.

The case for shipping using hydrogen-fuelled gas turbines is overwhelming. Australia's wish to use its coal resources to fuel the hydrolysis of water to make hydrogen which they can then export by ship to another country is mind-boggling; especially when they have thousands of square miles of emptiness and sunshine to equip with solar panels! One would think that with temperatures of 50° Centigrade and huge firestorms, they would have got the message!

The Law of Cosmic Equilibrium

If we look at how life on Earth has developed as a microstate, it is much like the way in which the extropy of 'order' consumes energy until a 'crisis point' is reached. Many species including our own have evolved into possessing large physical dimensions 'hosting' numerous much smaller organisms. The species known as Homo sapiens has now become so great in number that its consumption of energy is upsetting Earth's stability. This is contrary to the universal force of Nature, Cosmic Equilibrium. It goes without saying that this course of action, given the alternatives of renewable energy and conservation, is foolish to say the least. The Laws of Thermodynamics state that matter/energy can neither be created nor destroyed, although it may be rearranged or its entities changed in form. For example, in chemical reactions, the mass/energy of the chemical components before a reaction is equal to the mass/energy of the components after the reaction. Thus, during any chemical reaction or low-energy thermodynamic process in an isolated system, the total mass of the reactants, or starting materials, must be equal to the mass of the products. Despite the Cosmos being infinite, it acts as if is a 'closed' (isolated) system.

The state of Singularity cannot be described in finite terms since time, space and relativity which relate to the material world may well be illusory. The terms extropy and entropy describe states of 'open' or 'closed' systems. If a system is transferring energy into its surroundings due to exothermic reactions like on the surface of the Sun, the system is said to be extropic and the microstate benefiting from said energy is entropic. A forest-clad planet for instance, which absorbs energy from the sun due to endothermic reactions like photosynthesis can be said to be entropic and the system providing it with that energy is said to be extropic since its energy is depleting. Human beings continuously exchange energy with one another even by just 'shaking hands'. The direction of energy flow is partly based on their relative temperatures or electromagnetic 'charges'. We all know how difficult it is to come to terms with the 'emptiness' we feel when someone leaves home; for instance a child starting at university. At such times we can feel like 'opening the windows' to let new energy into the house. There's a scientific basis for such feelings if one thinks about it seriously. Both a source and a recipient of energy will have disappeared. A tangible and discernible change in the entropy/extropy balance of the house within its physical boundaries will have occurred. You can 'feel' it. Some of us 'abound' with energy, others may be lethargic with little energy to donate; we all play a part in maintaining the Cosmic Force of Equilibrium. The overall Cosmos is a balanced 'closed' system. Areas within it may be either entropic or extropic and states can exist in sizes ranging from the microscopic right up to the only 'closed' system of Equilibrium, the Cosmos. On Earth, surface effects vary enormously between two conditions, entropic and extropic. Overall, as an 'open' system it is absorbing a great deal more energy than it emits partly to do with the 'greenhouse' effect. This is the result of excessive surface reactions such as the combustion of carboniferous fuels. Balance, even locally, has now been foregone and sooner or later the Earth will become too hot for most species to survive on. Mini- and micro-cycles repeat themselves endlessly in every mini- and micro-state of the Cosmos. We would do well to mimic the way the Cosmos achieves its overall equilibrium. For billions of years apart from the odd super-volcano and meteor strike, Earth has done so. In just 200 years we have colossally disrupted the status quo. Up until now, our cyclic system has varied greatly but only over lengths of time great enough to allow species to adapt at a reasonable pace. The microstates of the Cosmos from the largest galaxies down to the individual microstate of every member of every species, don't often behave cataclysmically. The processes in the Cosmos to achieve equilibrium take place continuously by 'recycling' matter in numerous 'black holes'.

Perception, Objectivity, Subjectivity

This is such a popular subject that I'm surprised we haven't yet realised that we have no proof whatsoever that what we observe is coincident with reality, whatever our individual perceptions of it are. This isn't a philosophical supposition, it's a fact. We need to be more objective if we are to solve this important dilemma.

We assume that our perceptions of inorganic objects are reasonably accurate since they are usually common to the majority of us. A chair is a chair, a ball is a ball and the sun is the sun. Because our perceptions of objects are usually identical (it would be chaotic if they weren't!), we automatically assume that our senses are receiving the same inputs as those of everyone else. However, that doesn't necessarily mean that commonly sensed inputs are themselves 'uncontaminated'. If the information received by our senses has been corrupted en route, then we are being misled in important ways.

Plato described the human plight as like being trapped in a 'cave' with only 'shadows' within it as a guide to the reality 'outside'. We try hard to interpret the 'shadows' for clues but then according to Wittgenstein, we find ourselves in a 'fly bottle' of alternative perceptions. Surely, when we find ourselves in 'caves', shouldn't our first steps be to confirm that the 'shadows' themselves are reliable? Otherwise, it's a case of 'the blind leading the blind'.

We have evolved in ways by which we can experience the world we live in. Our principal method of communication is by light and electromagnetism. It's what we are constructed from. It travels at about 180,600 miles per second. Just as well since a good lip-reader armed with a powerful telescope will always be more up-to-date with fellow humans events than waiting for sound travelling at only 350 metres a second to reach him via their megaphones. Reality therefore, is a function of the speed of communication.

Light is far from perfect as a means of communication. The information it brings to us when we look out into space is always out-of-date by amounts of time varying from just a few seconds to 400 billion years depending on which part of the sky we point our telescopes. Even the Sun goes down 8 minutes before it gets dark here. Amazingly, we treat stellar objects as being present and real but the truth really is that we have no idea whether they exist!

There is a reality out there but the falseness we 'sense' of it is directly proportional to distance without it seems, taking account of this fact we convert these senses into false assumptions. We aren't alone in this quandary. For instance, a bat experiences the reality of a moth only when it's in his mouth. Until then, it has existed only as an echo, a 'sense' in its eardrums. Its brain turns this into an appetising perception deducing its size, speed and direction providing the bat with a course to navigate to intercept that of the moth. This illustration demonstrates that a bat's sense and perception is like our own, dependent on both our distance from, and the speed of communication with an event or object. In fact, a fellow bat could already have consumed the 'real' one.

    magine that you had the opportunity of travelling at an infinite speed to see what the reality of the Universe really looks like. You borrow this new invention, a 'time machine' and aim it at an interesting place full of stellar objects about 50 billion light-years from Earth and press the switch. Surprise, surprise, you find nothing there but space! Perplexed, you recapitulate and head back to Earth. You then rewind, re-target, and reset the machine to visit an area about the same distance from Earth but completely empty. This time you arrive at a place full of galaxies, planets and stars!

    You soon realise this isn't time travel. You are merely looking at what already exists but here on Earth you would have had to wait 50 billion years to experience it sensually! From wherever you view the Universe, you will see only out of date versions. The farther away an object is, the more 'ancient' is your sense of it. Wherever you are in the Universe, you will only experience the reality of your 'locality'. Universal reality would comprise everything being able to communicate with everything else at infinite speed but not necessarily able to participate in it.

I have discussed the difficulties endemic in our attempts to view objects at vast distances 'realistically'. However, even attempting to play chess with someone only as far away as Jupiter would be impossibly difficult. It would take 86 Earth minutes to learn how your opponent had reacted to your last move and vice versa. A game could last for 7 days not counting thinking time and your opponent could have died 90 minutes before receiving your last move. And vice versa of course!

On Earth, the reality of what we sense from objects is relatively accurate since our 'carrier postman' light, radio waves and electromagnetism is fast enough. Also the distances involved between communicants are small. However, our attempts to utilise objectivity in our quests for knowledge falls into insignificance when overwhelmingly our lives are ruled not by objectivism but by our individual and collective subjectivitism. Despite huge genetically and structurally similar characteristics, there aren't two of us who are subjectively identical.

Computers are a guide to our failings since we appear to construct them in our own image. They have huge storage databases (brains), contain numerous programmes called apps to handle particular tasks in their random access memories (frontal lobes) and most importantly an operating system (pineal gland) to make decisions. Unless a computer possesses an operating system someone has designed for it, it is just a meaningless mass of data. Its manufacturer provides this with 'motives' and purposes. What makes it impossible for computers to 'take over the world' is that they can only ever be the subjects of mankind, incapable of subjectivity's pleasures, pains and motivations and therefore committed only to objectivity. This doesn't prevent them becoming the tools of humans with subjective plans however.

All organic and self-determining entities possess their own unique 'operating systems'. Our survival is largely influenced by both our inherited and genetic capabilities; memory-stored experiences are also a factor. But, like computers we are provided with 'rules' and 'obligations' and we acquire these during our early post-natal and pre-school environments. These oblige us to absorb and 'programme' into our brains exclusive subjective perceptions of the world.

This 'conditioning' of our operating systems contributes to our perception of the world and our fellow occupants. As youngsters, we have little choice. However, this is not only because it's usually the only one available but also because it's viewed by most as our only means of survival. These carefully constructed subjective views of 'reality' can stretch across generations and must be learned and obeyed at least until we reach the independence of adulthood and consider alternative realities. To all intents and purposes however, they are 'carved in stone'.

When we choose to adopt alternative versions of our subjective 'realities' by attempting to change how we perceive the world, we can find that these can only be accommodated by making huge changes in our environments. Changes in how we 'see' things can resonate adversely throughout our friends and relatives. Each of us needs some self-assurance and acceptance to survive and thus exist as effectively as possible. Unless we are brave therefore, we unconsciously surround ourselves with situations and experiences which support our self-images. More often, we merely 'soldier on' in adversity rather than run the risk of disturbing the lives of our current 'myth-sharers' whose subjective worlds would be threatened by our 'new ones'.

If we were able to perceive from our senses an objective view of the world, life would undoubtedly be simpler. Instead of which, we look at almost everything subjectively ignoring its objectivity. Show a film clip to a room full of people and you will get as many subjective descriptions as there were people watching it. If we look at the plight of the world concerning Climate Change objectively, we see it threatened by the way we treat it and know we must urgently change our ways. However, sensing the subject subjectively induces a variety of perceptions of how one is likely to be affected by it, plans to tackle it, how one might profit financially from it and even how to tackle the crisis personally. When a common plan is needed, every town, city and country is subjectively making up its own mind how to go about things.

Viewing the Universe in its reality would always have been possible if light had always travelled at an infinite speed. Then, one's sense of everything in it would be synchronistic and real rather than historic but since we evolved without reality, we probably wouldn't now exist. However, we could be exploring alternative ways of communication. Quantum particles such as electrons 'entangle', communicating instantly with one another over infinite distances.

Currently, it's not within our ability either to do so or employ electrons to do so on our behalf. Singularity is one way of explain synchronicity. It's neither an infinitesimal point nor an all-encompassing infinite space. Neither exists if communication is instant.

Hints are provided to us from time to time of ways in which it might be possible. For instance, Synchronicity is the term ascribed to inexplicable coincidences of events which lack causality. This was researched by Carl Jung. We share electrons with everything there is. We may have much more to learn about them. I prefer that option to setting off to explore a Cosmos that doesn't really exist as anything more than an enticing illusion from which I can only make misleading perceotions. In any case, in my present form it would be impossible to explore it objectively.

Turning to the subject of philosophy itself, although I have taken an interest in various philosophers, I have tried to be as objective as I can in assessing the value of what they pronounced. Those that made objectively huge advances are fairly easy to credit. For instance, Charles Darwin provided us with an incredible amount of factual objective knowledge which enabled us to develop hugely in the fields of genetics and anthropology.

Surely, we should be educating young people to think more objectively. It should be the general practice in society to first objectively determine what a problem(s) is before attempting to objectively solve it! Subjectivity gets us nowhere and merely keeps us in Plato's 'cave of existence'. His description of it is surely the truest analogy.

Proposal for Completely Environmental Method of Water Treatment

Background

Various electrocoagulation techniques using external power sources have been used to rid water of unwanted suspended matter. In effluents such as water treated by sewerage purification plants and destined to be used for human consumption, contaminants in the form of suspended particles which are too small to remove by filtration are flocculated by the process into larger filterable particles which aggregate so that their removal by settling and decanting can be effected.

Electrocoagulation can be used to remove compounds suspended in water which may be too valuable to lose or which might constitute an environmental hazard. Heavy metals for instance are extremely harmful. Valuable metals such as aluminium compounds can be removed from waste water effluent arising from manufacturing pure aluminium from the mineral bauxite.

Chemicals have been used as coagulants, sometimes with serious results. In Cornwall, U.K., aluminium sulphate was being used to precipitate suspended material from the public water supply and was accidentally used to excess causing a health risk to water supply customers.

Chemical coagulation, whereby polymers are added to the water to be treated has been used for decades but the addition of chemicals tends to increase the Total Dissolved Solids (T.D.S.) content of the water making it unfit for immediate re-use.

Chemical treatment methods can enable the removal of suspended metal oxides, colloidal solids and particles, and soluble inorganic pollutants from aqueous media by introducing highly charged polymeric metal hydroxide species. These species neutralize the electrostatic charges on suspended solids and oil droplets to facilitate agglomeration or coagulation and resultant separation from the aqueous phase. Chemical treatment prompts the precipitation of certain metals and salts but can complicate the problems by adding further reactions and materials to the fluid.

Electrocoagulation offers an alternative to the use of metal salts or polymers and polyelectrolyte additives for breaking stable emulsions and suspensions. Electrocoagulation generally avoids the use of chemicals and secondary pollution caused by the addition of chemical substances and has the added advantage of removing the smallest colloidal particles by flocculating them into larger particles subject to gravity and prior art methods of filtration.

Direct current is used to produce various electrochemical reactions. Electrolysis creates small particles which then become the centres for larger stable, insoluble complex metal ions. Halogen complexing causes metal ions to bind to chlorine ions thus separating pesticides et cetera from water.

Electricity-based coagulation removes contaminants which are generally more difficult to remove by filtration or chemical treatment systems such as emulsified oil, refractory organics, total petroleum hydrocarbons, suspended solids and heavy metals. After the coagulation process, a variety of different removal methods may be used including settlement followed by decantation of the clear liquid and also filtration and sedimentation to remove the enlarged particles from the flow of effluent. Because the size of the coagulated particles is much greater than those in the untreated flow, they are much easier to dispose of. Filters become much more effective since filter membranes are not blocked by very fine particles such as colloids and can be cleaned more easily by back-flushing.

According to a known method of electrocoagulation, a reactor is made up of an electrolytic cell with one anode and one cathode. When connected to an external power source, the anode material will electrochemically corrode due to oxidation, while the cathode will be subjected to passivation. During electrolysis, the positive side undergoes anodic reactions, while on the negative side, cathodic reactions are encountered.

Consumable metal plates, such as iron or aluminium, are usually used as sacrificial electrodes to continuously produce ions in the water. The released ions neutralize the charges of the particles and thereby initiate coagulation. The released ions remove undesirable contaminants either by chemical reaction and precipitation, or by causing the colloidal materials to coalesce, which can then be removed by flotation. In addition, as water containing colloidal particulates, oils, or other contaminants move through the applied electric field, there may be ionization, electrolysis, hydrolysis, and free-radical formation which can alter the physical and chemical properties of water and contaminants. As a result, the reactive and excited state causes contaminants to be released from the water and destroyed or made less soluble. Examples of such prior methods of electrocoagulation using anodes and cathodes with an external power source are described, for example at:

https://en.wikipedia.org/wiki/Electrocoagulation;
http://www.gerberpumps.com/electrocoagulation-technology.html

There exists a problem for removing of suspended materials from fluids by coagulation that avoids the disadvantages of the known techniques described above. Reflecting on the actual chemical reaction of aluminium sulphate for example, one of the most widely-used coagulants, the reaction of that substance with water is exothermic. It needs therefore no extra supply of energy with which to form aluminium hydroxide which is relatively insoluble and acts as a flocculent which sweeps up colloidal impurities with it providing the water flow with heat. Provided only that the mixing process is efficient this will take place. It's reasonable to suppose that a low raw water temperature may help the process and a higher temperature may slow it.

Using calcium bicarbonate as a coagulant is quite a different affair since to dissociate calcium bicarbonate into its constituents, carbon dioxide, water and calcium carbonate is an endothermic reaction and thus needs heat to bring it about.

Recent research since my first patent in the early 1900's has demonstrated that the cavitation or vigorous agitation of water creates microscopically local temperatures of several thousand degrees Celsius as miniscule 'bubbles' of vacuum collapse violently upon themselves. It's my opinion and belief that this process not only adds electrons in the form of heat increasing the general entropy of the water but more importantly creating nucleic particles of calcium carbonate as aragonite. These 'lay the ground' for further deposition of calcium carbonate during the passage of the raw water through the various stages of treatment in a typical water works.

Because the cavitation/agitation takes place in the presence of a relatively large surface area of PTFE (the most strongly negatively charging material known), the precipitated particles of calcium carbonate become positively charged and neutralise the negatively charged colloids of impurity 'cancelling out' the Brownian Movement of their mutual charge.

In the early 1990's I successfully invented a method of coagulating suspended particles in a flowing stream of water by incorporating the use of a dielectric material in a flowing stream of contaminated water. It became hugely successful worldwide and is still being marketed worldwide principally to reduce the effects of hard water scale in various heating and pipework constructions such as cooling towers, heat exchangers, etc.,

Some interesting results have shown that merely by following the mixing and agitation procedures set out in the standard Glass Jar Tests without utilising either my equipment or using a chemical coagulant such as aluminium sulphate or ferric salts resulted in very encouraging analyses. These compared favourably with those obtained by using my own invention and/or chemical coagulants. I am concerned that the water industry may be missing out on a hugely clean and simple way in which to purify water without polluting the world with billions of tonnes of aluminium sulphate and similar chemical coagulants.

Undoubtedly, my previous patent (now expired) works well. I believe that because it was aimed principally at the scale, corrosion problems common in the water industry, its use in the treatment of water to remove colloidal suspended particles was not examined to any great extent.

I now believe that the effect of causing the nucleation of microscopic particles of precipitated calcium carbonate by destabilising the bicarbonate molecule has escaped close scrutiny. These microscopic particles grow in size and coagulate further to capture the colloids of impurities eventually to sink as sediment in a gently flowing or stationary stream of water or to be removed by relatively unsophisticated filtration techniques.

.

Energy & Power

I recently read a section in a children's encyclopaedia' describing how when one object is at a higher temperature than another, energy will flow between them until their temperatures are equal. If only we could always use such simple language with which to understand Energy so easily. Life isn't that simple! Since our whole existence has relied on energy, its production and utilisation, perhaps the Cosmos is a good subject with which to start discussing its importance.

Einstein encouragingly suggested that "*philosophical thinking makes for "the distinction between a mere artisan or specialist and a real seeker after truth"*. He and Charles Darwin were regarded as both philosophers and scientists; they have been at the forefront of knowledge about ourselves and how and why we exist. Having once been a research chemist and in the last 10 years having read about the lives of some of the great philosophers, I have come down strongly in favour of Plato's suggestion that to access 'truth' we need to use 'reason' to get out of the 'cave of ignorance' he suggested we all live in.

The 'information' our 'senses' perceive about the Cosmos is out of date by up to billions of years. Even the light from the centre of our own galaxy takes 25,000 years to reach us. The sun goes down 8 minutes before it disappears here on Earth. Due to the refraction of starlight as it passes through our atmosphere stars falsely appear to be clustered towards the middle of the sky and despite that knowledge little attention has been paid to the fact discovered 100 years ago that gravity bends light. The result is that when light and its energy pass close to large stellar objects such as stars, it changes course due to their gravitational effects on light's path. This can be illustrated by poking a pencil in a bowl of water. The pencil will appear to bend as it enters its surface.

Not only are we somewhat ignorant of what is really out there, we don't know for certain where light from sighted objects originated! Despite these intangibles, scientists persist in describing various objects in the Universe in the present tense treating what they can see as the 'truth' when it's anything but the 'truth'. They don't fool me. As Plato suggested, all of the intelligence and scientific knowledge available will still leave us with only 'informed guesses' as to the reality of what we 'sense' and even whether it still exists. We can only 'reason'.

Contrasting Philosophy which I thought until recently was using one's brain to 'reason things out' with Science, raises some problems since those professing to be philosophers tend to believe everything they are told by scientists despite the drawbacks I have described. The subject I have chosen to discuss is quite a testing one if not enigmatic since it dares me to comment on many of the things we all take for granted. This is an activity philosophers could might indulge in more frequently but they seem content to dwell in a world of infinite interpretations of papers written very often either in a bygone age, under vastly different circumstances or both by their original authors.

Have we become preoccupied with studying Philosophy as a historical progression of texts to be assimilated by its students rather than using philosophy as a means to subject our suppositions to 'reasoning'? This might produce ideas more consistent with the world we actually live in. We turn out thousands of scholars of Philosophy schooled in the interpretations of various philosophers from Aristotle to Zubiri. Have we become receptacles of past philosophies, constantly modifying them to suit our senses of reality and then regurgitating them as new ones? As a group, do they resist originality by merely reframing old ideas stamped with their own senses of reality?

Energy is currently topmost in our minds because of Global Warming. We aren't producing 'green' energy to replace carboniferous fuels quickly enough. This global problem offers us an opportunity to use our fundamental philosophising abilities to find solutions. Politicians are in the main preoccupied with their own agendas and seem unwilling to be guided by rational thought. China's rulers thousands of years ago usually had philosophers at their sides to help in their decision-making. Today's substitutes, their 'political advisers' don't help our plight.

Scientists have arguably got us into the Climate Change mess we are currently in. Even 100 years ago, simple arithmetic would have indicated the importance of energy, in particular how we source and store it. I'll illustrate their abysmally poor ways of 'reasoning'. Four years ago, I described in a published work how the Universe is governed by what I term the Universal Law of Equilibrium. I attempted to explain how it governs itself. It's basically as simple as shown in the children's book; energy will always flow from places and objects high in energy content to those with less. When we shake hands with one another the slightest difference in the temperatures of our hands will cause an immediate transfer of heat in the form of electrons.

It's not quite that simple but Gravity plays a huge part in maintaining Equilibrium. If the Law of Equilibrium didn't exist, neither would the Universe since we know that the infinite mass and gravity of 'black holes' would by now have contracted everything into a single point and given theology an opportunity to believe   Creation Theory when it unbelievably exploded to create the Universe as we know it. Conversely, the unhindered conversion of mass into infinite energy would have dissipated everything into an infinite space. This conundrum has baffled scientists for centuries resulting in what I consider to be extraordinarily ridiculous theories like the 'Big Bang' and an 'expanding Universe' to explain a velocity of expansion exceeding that of light. Whether God exists or not is a personal choice but if He does I prefer to think He dwells in an infinitely changing, infinitely huge Universe maintained infinitely in a state of overall equilibrium.

In my opinion, black holes are merely places where infinite accumulations of matter (mass) create infinite gravity. This converts it back into energy spitting it back to fuel the Universe with power to recreate matter somewhere everywhere. Black holes are effectively recycling centres maintaining universal Equilibrium. The infinitesimal particles of energy they project move at an infinite velocity.

The German philosopher **Leibniz** talked about "monads" as the indivisible soul-like particles that are the ultimate elements of the universe. I prefer to think of them as particles so tiny they cannot be identified by us. Sacrificing their mass for velocity is very effective since E = mass x the square of velocity and mass-free objects can escape the clutches of black holes when light clearly cannot. Incidentally, if you divide a substance of matter into two parts an infinite number of times there will always be an infinite number of particles possessing infinitesimal energy but moving at an infinite velocity.

The First Law of Thermodynamics states that heat is a form of energy, and thermodynamic processes are therefore subject to the principle of Conservation of Energy. This means that energy and matter can transform in many ways but cannot be destroyed. Energy resulting from burning carbon for instance can continuously change the form of its recipients from ice to water to steam to ice but energy can never be destroyed.

These facts are generally expressed in complicated mathematical ways distracting us from seeking philosophised descriptions of how everything in the Universe functions according to its relationship with everything else. Einstein rightly found ways to describe this with his Theory of Relativity. Put simply, it means that any entity can be described only in relation to everything surrounding it. Take two cars for instance, one of which is overtaking the other. One could be travelling at 70 mph, the other at 69 mph. They could bump into each other and their drivers might hardly notice it both with relative speeds of 1 mph. Their cars' energy levels are fairly similar so they pose no danger to one another but if one driver released some of his car's energy by braking to a standstill they might as well be on different planets since they would have become physically hugely unrelated!

The Universe has its own means by which to continue indefinitely in a state of overall Equilibrium. Within its infinity, there are infinite ecosystems of every size from blood cells to planetary bodies and galaxies all at various stages of development and destruction. Each one is obliged to follow the 'general rule' of Equilibrium or cease to exist. Earth is no different. We can choose if we wish to lead lives seeking personal, physical and social Equilibrium with one another and cooperatively utilise the energy of the sun as sunlight which is free and abundant. This would be in sharp but a 'well-reasoned' strategy to adopt instead of treating Climate Change as yet another means to profit by adding mass or energy to our personal stores in its many forms from bank accounts, pyramids, bitcoins and cellars full of Old Masters.

We need to steer ourselves towards global Equilibrium. Neither the entropic chaos due to Earth's overabundance of energy, nor hastening its extinction as surplus mass in a black hole are attractive goals. The process is only beginning to be painful. It can only get worse. With Equilibrium guiding our lives we wouldn't need Commandments to define 'mortal sins'. They wouldn't occur. We could start straight away by dealing with inequality! Looked at from distant space by aliens, this must seem to them like hell and wisely left to its own devices!

The Big Bang Theory Banished

If we continue to believe in the Big Bang theory after reading what I have had to say in this collection of articles, we will be wasting the short space of time left for us to stop our path of self-destruction.

Homo sapiens, whom one BBC anthology researcher described recently as always having been quite 'mad', have inflicted greed and exploitation on everything that existed since we emerged from Africa. We subsumed our predecessors, the Neanderthals and Denisovans who lived gentler, more sustainable lives on the planet for a much longer period of time than us.

We have consistently treated one another and the Earth disrespectfully in our interminable search for wealth and power. As nations, we have pillaged less-developed communities enslaving their long-term residents to satisfy our own insatiable needs.

'Fake news 'about the Cosmos is constantly weaved in a search for scientific status. I hope to demonstrate to you that we possess no 'actual' knowledge at all about the Universe; we live our lives only on the basis of our 'perceptions'. We need to stop draining our resources on panaceas of hope by dreams of 'escaping to new far off planets' and concentrate on the here and now.

Introduction

From philosophy.com on Plato, *"The real is divided into two parts: first the physical world accessible to the senses, the real immediate source of error and illusion, the other the intelligible world accessible to reason alone, of ideas and truth. Combining reality and truth, Plato condemns the world of sense. The horse is not the truth, only the idea of a horse is true. Thus, the "Cave" means the material world, whose wise-philosopher has to divert to the world of ideas. Access to the Truth through contemplation, the exercise is to make use of his reason."*

The great scientist Einstein was quoted as saying that "The true sign of intelligence is not knowledge but imagination", and "The intuitive mind is a sacred gift and the rational mind is a faithful servant. We have created a society that honours the servant and has forgotten the gift."

Taking him at his word therefore and assuming also that Plato was wise, I plan only to visualise the 'truth' about the Cosmos in order to rationalise my way out of

Plato's 'cave'. The *truth* is actually quite inaccessible; I'll explain why!

Scientific theories should only be deemed correct if sufficient basic mathematical parameters have been first established with which to draw conclusions. We can reliably employ mathematics and the Laws of Thermodynamics to aid our research but only up to a point. Much theorising about the nature of the Universe appears to have avoided this discipline. Assumptions have been made that there is a finite relationship between certain variables by estimating their values without substantiating them with at least one incontrovertible constant.

Trigonometry can only provide ratios and relative associations between variables. Without knowing at least one *actual* value, others cannot be deduced. Put simply, you need to know the values of two angles of a triangle to establish the third. Knowing all three angles does not provide its area. The lengths of two sides of a triangle will not provide the length of its third side, its angles or its area.

The Big Bang Theory has been fabricated on the basis of *apparent* galactic relationships. At various times, an association between red light, the speed of recession of a terrestrial object and the distance of the object emitting its light has been assumed. These claims have supported the theory that the Universe (space) is accelerating outwards. By 'inverse extrapolation', we have then dated the Universe's genesis!

If this isn't being stuck in a 'cave', what is? I will make only basic mathematical equations to clarify a subject which has become shrouded in obscure scientific theorems. I hope this paper proves to be the 'simple' way to change our understanding of the Universe which we appear to be searching for which will reshape physics towards the 'rational' rather than the 'sensational'.

Hopefully perhaps, it will attract the scientific establishment away from seeking answers which do not exist to concentrating their activities on the 'here and now'. We are under threat of extinction and our skills would be better applied to our real problems. My intention is only to bring some common sense into what has become a long-standing debate.

Abstract

Copernican Heliocentrism is the name given to the astronomical model developed by Nicolaus Copernicus and published in 1543. This positioned the Sun near the centre of the Universe, motionless, with Earth and the other planets orbiting around it in circular paths.

Following that view, a model of the Universe as a static state of equilibrium prevailed until about 100 years ago when widely attributable to Edwin Hubble, the notion of the universe expanding at a *calculable* rate was derived from a set of general relativity equations produced in 1922 and now known as the Friedmann equations. These showed that the universe 'might' expand, and presented the expansion speed if this was the case. Georges Lemaître, a Catholic Priest stated in a 1927 article that the universe 'might' be expanding, observing the proportionality between recessional velocity of and distance to, distant bodies.

With further encouraging work by Hubble, it was 'established' that there was a *linear relationship between redshift and distance*. From this, it was assumed that the distance of an object from Earth could be by measured by light's red shift. I quote from Wikipedia:

"In physics, redshift is a phenomenon where electromagnetic radiation (such as light) from an object undergoes an increase in wavelength. Whether or not the radiation is visible, "redshift" means an increase in wavelength, equivalent to a decrease in wave frequency and photon energy in accordance with the wave and quantum theories of light.

There are three 'stated' causes of redshifts in astronomy and cosmology:

1. *Objects move apart (or closer together) in space. This is an example of the Doppler Effect.*

2. *Space itself expanding, causing objects to become separated without changing their positions in space. This is known as cosmological redshift. All sufficiently distant light sources (generally more than a few million light years away) show redshift corresponding to the rate of increase in their distance from Earth, known as Hubble's Law."*

3. *Gravitational red shift which is a relativistic effect observed due to strong gravitational fields which distort space time and exert a force on light and other particles."*

My view is that objects only *appear* to move apart or closer together in space and red shift only *appears* to correspond with the rate of increase of objects' <u>radial</u> distance from Earth. I will deal specifically with points 1 and 2 above later when I have fully explained why objects at great distances not only *appear* to separate but also *appear* to be separating at an accelerating rate.

Since the 1920's and largely as a result of red shift measurements, the Big Bang Theory has been consistently supported by the scientific sector. Every modern student of astrophysics, physics and associated subjects has been 'inoculated' from birth with the following beliefs summarised by Wikipedia.

> *"Hubble's law, also known as the Hubble–Lemaître law, is the observation in physical cosmology that:*
>
> 1. *Objects observed in deep space—extragalactic space, 10 megaparsecs (Mpc) or more—are found to have a redshift, interpreted as a relative velocity away from Earth;*
>
> 2. *This Doppler shift-measured velocity of various galaxies receding from the Earth is approximately proportional to their distance from the Earth for galaxies up to a few hundred megaparsecs away.*
>
> 3. *Hubble's law is considered the first observational basis for the expansion of the universe and today serves as one of the pieces of evidence most often cited in support of the Big Bang model. The motion of astronomical objects due solely to this expansion is known as the Hubble flow."*

An enigmatic point was made by Maria Tenning in 'Sky and Telescope', 2014 which rather unscientifically deduces one possibility from another!

"The age of the universe is approximately 13.77 billion years. This age is calculated by measuring the distances and radial velocities of other galaxies, most of which are flying away from our own at speeds proportional to their distances. Using the current expansion rate of the universe, we can imagine "rewinding" the universe to the point where everything was contained in a singularity, and calculate how much time must have passed between that moment (the Big Bang) and the present."

From the assumption that red shift measurements have a linear relationship with objects' distances from Earth, the Doppler Effect is supposed to also associate red shift with recessional velocity. How can red shift be indicative of two variables; distance and velocity?

Loudspeakers sometimes emit ear-piercing shrieks of noise. This is caused by what is known as a 'feed-back loop'. Put simply, it occurs when a microphone picks up sound from the loudspeakers in the auditorium which is then amplified before finding its way back to the microphone. This 'loop' of magnification and sound emission quickly produces a cacophonous noise quite unrelated to the original voice of the announcer. Have we allowed ourselves to get caught up in a Big Bang 'loop' by concentrating on *appearances* instead of 'reasoning' the 'truth'?

A Universe expanding exponentially at speeds greater than the speed of light would obviously need a huge amount of energy. Its 'source' has been 'surmised' as 'dark energy' in the form of the 'Boson' particle named after its 'inventor'. This was further theorised as particles which possess enormous mass compared to other sub-atomic particles. Despite the construction of the Hadron Collider at Cern and subsequent research over many years, the 'Boson' particle remains theoretical although its *'discovery'* merited a Nobel Prize.

A huge amount of mathematical *theory*, not to mention the cost of over $100 billion has been expended in order to support the following surmises; that red shift in light from distant objects is proportional to their rates of recession from the Earth, one another and objects' distances from Earth thus deriving the age of the Universe. Refreshingly, this 'ray of light' has just appeared in 'live.science.com':

'A crisis in physics may have just gotten deeper. By looking at how the light from distant bright objects is bent, researchers have increased the

discrepancy between different methods for calculating the expansion rate of the universe.

"The measurements are consistent with indicating a crisis in cosmology," Geoff Chih-Fan Chen, a cosmologist at the University of California, Davis, said here during a news briefing on Wednesday (Jan. 8) at the 235th meeting of the American Astronomical Society in Honolulu'.

Discussion

From Wikipedia:

'For galaxies more distant than the Local Group and the nearby Virgo Cluster, but within a thousand megaparsecs or so, the redshift is approximately proportional to the galaxy's distance. This correlation was first observed by Edwin Hubble and has come to be known as Hubble's law. Vesto Slipher was the first to discover galactic redshifts, in about the year 1912; while Hubble correlated Slipher's measurements with distances he measured by other means to formulate his Law. In the widely accepted cosmological model based on general relativity, redshift is mainly a result of the expansion of space: this means that the farther away a galaxy is from us, the more the space

has expanded in the time since the light left that galaxy, so the more the light has been stretched, the more redshifted the light is, and so the faster it appears to be moving away from us. Hubble's law follows in part from the Copernican principle. Because it is usually not known how luminous objects are, measuring the redshift is easier than more direct distance measurements, so redshift is sometimes in practice converted to a crude distance measurement using Hubble's law.'

I will demonstrate why Red shift gives an observer only an *apparent* measure of the radial distance covered by the light from its source. It provides instead, an indication of the *actual* distance light has taken en route to an object's observer!

From the website 'Cosmos':

"Einstein's theory of general relativity predicts that the wavelength of electromagnetic radiation will lengthen as it climbs out of a gravitational well. Photons must expend energy to escape, but at the same time must always travel at the speed of light, so this energy must be lost through a change of frequency rather than a change in speed. If the energy of the photon decreases, the frequency also decreases.

This corresponds to an increase in the wavelength of the photon, or a shift to the red end of the electromagnetic spectrum – hence the name: gravitational redshift. This effect was confirmed in laboratory experiments conducted in the 1960s.........For radiation emitted in a strong gravitational field, such as from the surface of a neutron star or close to the event horizon of a black hole, the gravitational redshift can be very large................"

Ethan Siegel (Twitter) wrote:

"What happens to light when it passes near a large mass? Does it simply continue in a straight line, undeflected from its original path? Does it experience a force owing to the gravitational effects of the matter nearby? And if so, what is the magnitude of the force it experiences?

"These questions cut to the very heart of how gravity works. The year, 2019, marked the 100th anniversary of General Relativity's confirmation. Two independent teams undertook a successful expedition to measure the positions of stars near the limb of the Sun during the total solar eclipse of May 29, 1919. Through the highest-quality observations that technology permitted at the time, they determined whether that distant starlight was bent by the Sun's gravity, and by

how much. It was a result that shocked many, but Einstein already knew what the answer would be."

An important point to bear in mind is that when light (photons) fall into a gravitational field they gain energy and their wavelengths become blue shifted. Upon climbing out of a gravitational field, they lose energy and their wavelengths become red shifted.

This reveals an extremely important question. If the foregoing is true and was known when red shift was becoming 'cast in stone' 100 years ago as a measure of distance/recessional speed, why wasn't more attention paid to the fact that red shift can be caused by gravitational effects? The Hubble Constant appears to be only an attempt to establish a constant rate of expansion. Surely, the experiment described above should have 'thrown a spanner in the works' or at least rung some 'alarm bells'?

Red shift is apparently being used as a constant to deduce two interdependent variables, speed of recession and distance from Earth. Unless we can prove without doubt that the relationship between red shift and an object's velocity away from Earth and/or its relationship to the deflection of light caused by gravitational bodies en route we are in trouble. Any assumptions made without such assurances are suspect.

However,

1. Red shift appears to measure something.

2. Red shift can be due to either the recessional speed of a stellar object from an observer and/or the effect of gravitational bodies and background cosmic radiation encountered by light en route to an observer on Earth.

3. Red shift measurements from relatively close stellar objects in our own galaxy *appear* to be proportionally related to their distance calculated by traditional parallax means but red shift measurements from vastly distant galaxies indicate objects' 'acceleration' away from us with distance'.

With regard to point 1 above, Rob Jeffries' made a most interesting point in the online site, "@Physics" which rules out the use of parallax to measure huge distances. This casts great doubt on the use of physical measurements to determine galactic distances.

"The realms of applicability for redshift (as an indirect distance indicator that must be calibrated through the Hubble parameter) and parallax, which can only be applied to relatively nearby stars, do not overlap.

Parallax is a geometric method, limited only by the precision with which the parallax can be measured. At present it is limited to distances of hundreds of light years, though the Gaia astrometry satellite will soon extend this to thousands of light years.

Redshift as a (reliable) distance indicator can only be used well into the "Hubble flow", so that individual Galactic peculiar velocities become unimportant compared with the universal expansion. In practice this means tens of millions of lightyears at least. The technique cannot be used at all on local group galaxies or stars in our galaxy, since their motions are not due to the expansion of the universe.

You are correct that parallax is more accurate, but current levels of measurement precision are too low to extend the technique to distant galaxies.

One way of looking at this is to ask what parallax baseline would be needed to measure a parallax at 100 million light years with current measurement technology.

Gaia can measure stellar positions to about 10 microarcseconds. The parallax at 100 Mlyr is about $0.033b$ microarcseconds, where b is the parallax baseline in astronomical units (the Earth-Sun distance). To get a parallax measurement precise to 10% would require a *baseline of b~3000b~3000 au - i.e. larger than the solar system."*

An important thing to understand is that direct observations of distant stars and their separating distances only provide their *apparent* positions relative to one another. The distance light will have *apparently* travelled from an object to Earth. In fact, red shift will be a measure of its *actual* distance travelled from the object to Earth.

Endless mathematical equations have been used to justify the Big Bang theory which has unfortunately disenfranchised the non-scientific community from involvement with the subject. For that reason, I don't intend to follow the same route; I will take the path that Einstein used which was to 'picture' problems and their solutions. As the great man said *'Concern for man and his fate must always form the chief interest of all technical endeavours. Never forget this in the midst of your diagrams and equations.'*

I'm encouraged by that thought since a scientific journal published this month stated that the world was waiting for just one simple explanation to alter physics substantially. I am confident I can provide one.

I'm sure that by now you are aware that I have been discriminating not only between a celestial object's *apparent* position and its *actual* position but also between its *apparent* and its *actual* <u>distance travelled</u>.

Until we find a way of viewing the Universe which doesn't involve using light, we will be stuck where we are now; attempting to deduce what is out there from *apparent* information and not *actual information*. When we accept the inevitable fact that we cannot even remotely 'rely on our senses' as Plato suggested, we will continue to waste valuable time and resources 'chasing shadows'. Everything in our self-important lives particularly events which occur at great distances, is pure illusion.

An interval of time will always separate our sensual perceptions from the 'truth' of the 'intelligible' world. Because information by way of light or radio waves is conveyed at only 186,300 miles per second (the speed of light), on Earth only the tiniest interval of time separates our *'sensing'* an event from the *actual* event. With greatly increased distances, the time interval becomes significant. For instance, if the sun 'went out', our *'sense*s' wouldn't perceive that event until 8 minutes later when darkness would descend on us.

Consider another example; a bat creates high pitched sounds and can only *'sense'* a flying insect from the sounds echoing back from it. Information conveyed by sound in air travels at the relatively slow speed of about 760 mph, so a bat's *apparent 'sense'* of where an insect is could be about half a second 'out of date' with its *actual* position. The bat's brain therefore has to *'imagine'* its owner's 'flight path' and set a route for its owner to follow taking into account the interval between it emitting a sound and hearing its echo. By making continuous adjustments, the bat's flight intersects with that of the insect. Metaphorically, the *'reason'* or *'idea'* of the bat's brain has turned the *'illusion'* of its *'sense'* into *'truth'*.

You may think these scenarios rather tedious but please humour me for the time being. Their importance will soon become clear to you.

Imagine you are wearing a mask, goggles and wetsuit and sitting comfortably in the bottom of a large tank of water looking upwards at a torch which is shining down on you. Its light has passed into the water at an angle to its surface. This is termed the angle of incidence. Because of the refraction of the beam of light, it will bend downwards as it enters the water and the torch will thus *appear* to be higher than its *actual* position. Alternatively, before you next have a bath, poke a bamboo cane into the bathwater and it will *appear* to bend upwards as light passes through the surface of the water to you.

We know that the stick only *appears* to have bent. Light reflected from the stick you poked into the bath has bent downwards as it leaves the water into the much less dense air, giving you the idea that the stick has bent upwards. In each example, we have two 'states' to think about concerning the *apparent* stick and the *actual* stick and the *apparent* torch and the *actual* torch.

Bear in mind that because of light's refraction caused by a change in density of the medium, the *actual* distance light travels to reach your eyes from either the farthest point of the stick immersed in water or the torch shining down to a submerged you, will be significantly greater than if there was no water involved! That would be the *actual* distance travelled in each case.

When light is bent en route to your eyes, its first 'leg' of the journey, path A is from the end of the stick to the surface of the water and the second 'leg' of its journey, Path B is from the surface of the water to your eyes. These two separate 'legs' of the journey can be represented by two sides of a triangle, A and B. (A + B) will be the *actual* distance the light will travel to an observer. C, the third side of the triangle, will represent the path light would follow directly to your eyes from the end of the stick if there was no water in the bath to refract it. A + B will always be greater than C. (A + B) is the *apparent* distance of the object from your eyes measured by red shift when it is *actually* a measure of the total distance covered. Following any deviation, the distance travelled indirectly to an observer will always be greater than the actual distance it would travels without an intervening deviation. <u>More importantly, the red shift would be increased as a measure of the increased distance travelled.</u>

Without your being able to accurately calculate exactly how much the water has bent light you don't really know where the far end of a 'bent' stick *actually* is! If there was a nice trout in a pond, you were hungry and had a spear in your hand, you would have to guess where to aim knowing it would be some distance below where it *appeared* to be and hope for good luck.

As I explained earlier, when light from a star reaches our atmosphere, it will be refracted downwards in exactly the same way as when it hit the surface of the water in the imaginary tank of water. As light from a star hits Earth's atmosphere, it will be bent by refraction. Since our atmosphere is denser than the vacuum of space, when light gets bent it loses energy and this will increase its light's red shift.

As starlight passes through successive layers of Earth's atmosphere, each one a bit denser than the one before it, it will be refracted more and more. Its path through the atmosphere will be bent. If the medium through which light passes changes in density, it will cause light to bend successively.

The use of parallax to measure objects' separation is based on their *apparent* positions in the sky and not their *actual* positions. Since the light to measure parallax will be 'identical' to the light used to measure red shift, they can hardly be used to provide checks on each other. Neither will be the 'truth'.

Sitting in a tank of water and looking upwards, you don't know the *actual* position of the torch. You will only have a rough idea of where it is. Similarly, when you look upwards from the surface of the Earth at a star, you won't actually know for certain where it *actually* is. What's more, you won't know its *actual* distance. You will know fairly accurately however, how far its light has travelled to reach you because you can measure its red shift. This will indicate how much energy it has expended.

Incidentally, because of atmospheric refraction, stars *appear* to group themselves more towards the centre of the night sky. They are *actually* much more spread out!

It's hard to imagine that light reaching Earth from objects billions of light years away could possibly reach us without encountering some gravitational bodies en route such as stars, galaxies and black holes. We have no way of knowing how many encounters that light from a particular star will have on its way to us. Thus we have no idea of how many gravitational deflections will have been incurred on its path.

Light from every luminous body we can see in the sky will have had its journey length increased by an unknown amount. Each deviation will not only increase light's distance travelled to reach us but will have an exponential effect on its overall journey. Put simply, successive increments of 1^0 of deviation (angle of deviation) will result in larger and larger distances light must travel to reach any particular radial distance from its source.

Let me describe the term exponential in relation to global warming. Quite unbelievably, we assumed that the increase in carbon dioxide in the Earth's atmosphere would increase linearly with time. This avoided other considerations. For instance, when previously ice-covered and reflective surfaces started to melt, this has exposed heat-absorbent land adding to the temperature of the Earth which then causes more ice to melt.

Currently, huge tracts of land in Australia are burning because of the planet over-heating. This is destroying forests and their ability to absorb carbon dioxide causing even more heat absorption, more increase in temperature and so on...... This is an exponential effect in response to a stimulus, not a linear response.

When gravity causes deviations in light's route to Earth, their effects are not linear, they are exponential. The greater the degree of deviation (angle of incidenc) of light caused by gravitational bodies, there will exponential increases in *actual* distance travelled at the expense of radial (*actual*) distances achieved thus an exponential increase in red shift in both the time and distance covered exposing light to an increased risk of deviated by gravity.

Snell's work in the 16th century demonstrated that the angle of incidence of light passing from one medium such as space to air or from air to water is directly proportional to its wavelength and thus its red shift. I think you will now understand why some of the assumptions made about the Universe 100 years ago looking upwards through the atmosphere would have been subject to significant error.

An observer on Earth viewing two separate objects will know that refraction effects caused by our atmosphere will mean that viewed simultaneously they will *appear* to be closer together than they *actually* are. Below is a drawing from quora.com which illustrates the point.

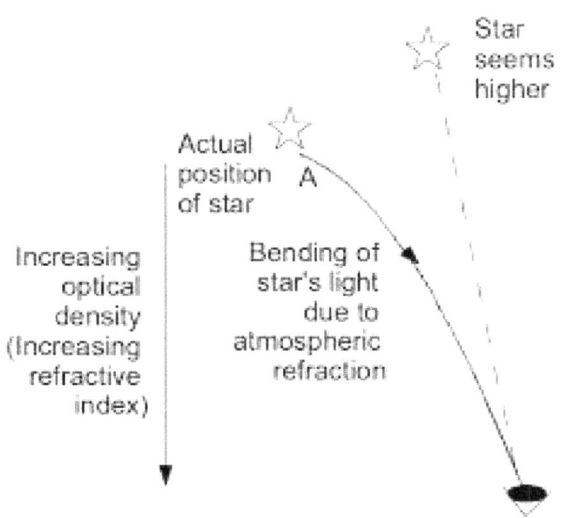

Star
seems
higher

Actual
position   A
of star

Increasing
optical
density
(Increasing
refractive
index)

Bending of
star's light
due to
atmospheric
refraction

This throws doubt on the conclusions made by early astronomers. One hundred years ago, when the Big Bang Theory was in its infancy, they would have been aware of the effects of refraction and would have endeavoured to obviate these from their calculations by trying as far as possible to focus on objects only when they were directly overhead. The angle of incidence of the light with Earth's atmosphere would then be as small as possible and would therefore cause the least refraction. However, there are only so many stars and galaxies that can be simultaneously observed in this way. In any case, their *apparent* positions would also be used to measure parallax so one incorrect assumption would merely cause another.

Explaining the problem

There are distinct similarities between the behaviour of light when it travels through a medium such as Earth's atmosphere or water and when it travels through space. In the latter instance, light is not refracted because in the main, the density of space is fairly constant; this is generalised as cosmic radiation and I believe the Hubble Constant was initially devised to account for this in its calculations. This might not necessarily bend light since its effect on light would be homogeneous and therefore evenly applied from every direction. It would more closely resemble a 'thin pea soup'.

Light travelling from A to any point B on a radius C from point A would be bent when passing close to a large gravitational body. Such a deviation{s} would not only lengthen the *actual* distance light would travel giving a false *radial* distance from its source; it would lengthen the time taken on the journey. Any increase in light's journey distance and time would impose an increased duration of exposure to not only energy loss caused by the Hubble Constant but also energy losses caused by the proximity of gravitational bodies en route. I would like to reiterate the point made earlier that light gains energy when it is pulled towards a gravitational body and loses it when pulling out of one. The net loss may be very small but will be cumulative nevertheless.

On 9th January 2020, the following was part of an article published by "Livescience.com":

"A crisis in physics may have just gotten deeper. By looking at how the light from distant bright objects is bent, researchers have increased the discrepancy between different methods for calculating the expansion rate of the universe.

"The measurements are consistent with indicating a crisis in cosmology," Geoff Chih-Fan Chen, a cosmologist at the University of California, Davis, said here during a news briefing on Wednesday (Jan. 8) at the 235th meeting of the American Astronomical Society in Honolulu."

I wonder how we have taken so long to admit there are serious doubts surrounding the Big Bang Theory; the 'emperors' clothes'?

I have elaborated on what happens to light as it passes through Earth's atmosphere as an analogy in order to demonstrate what happens to light as it travels to us from billions of light years away. I have emphasised how powerful the tiniest gravitational effect can be when multiplied over time. If light from a star, galaxy or other luminous body *actually* 100 billion light years away passed close enough to a gravitational body (for example a black hole) and was deflected by gravity early in its journey by only $1\sec^0$ (1/3600 of one degree), and continued towards Earth without any further impediments, its *actual* position would be literally about 480,000 light years from its *apparent* position. That's an error of about 20 times the width of our own galaxy.

This shows that we cannot determine very much at all about the Universe. We have no means of knowing how many deflections light will undergo on any journey it makes and will therefore have no idea of any objects' distance or place of origin in the sky. Any information carried to us by light originating from sources billions of light years away ago would be well 'past its sell-by date'. Perhaps you will now share my frustration with a scientific community which continually speaks about events in the Universe which happened long ago as 'taking place' or describes them in the 'present tense'. We don't have a clue of what is currently happening in the Universe. We can only 'reason'.

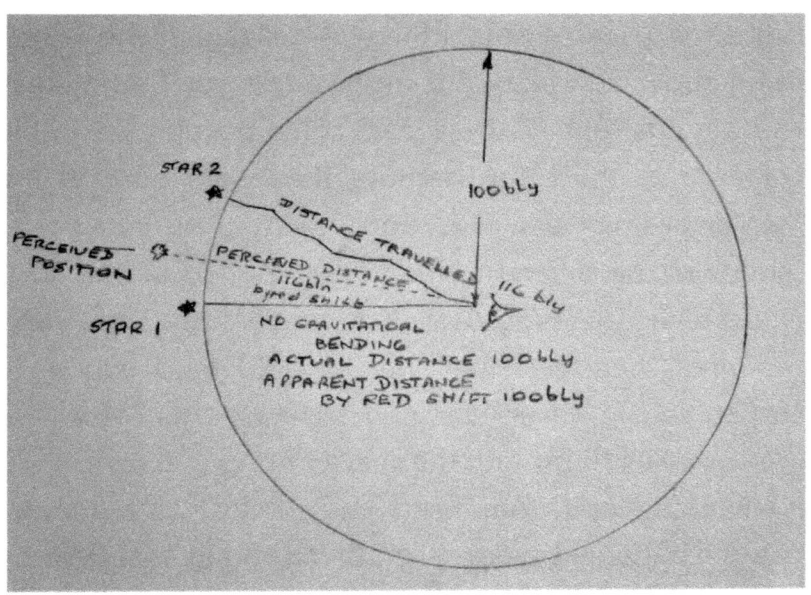

I have inserted above a rather crude diagram to illustrate the enigma we must overcome.

Two stars occupy different points on the circumference of a 'circle' with a radius of 100 billion light years (bly). As you will see light from star 1 quite extraordinarily meets up with no gravitational impediments at all on its journey to you at the centre of the 'circle' indicated by an eye. You whip out your red shift measuring device and expose it to the light from star 1 and it tells you that its distance is close to 100 bly.

You think, that's nice to know, I'll do the same for star 2. Lo and behold, this time red shift tells you that this one is 116 bly distant. Their respective luminosities tell you the same story. Why wouldn't they? The same light was used to obtain them! A colleague who knows a lot about how stars and suchlike tells you that there is no way that one of these stars could possibly be 16 bly older than the other because they were born in the same cosmic era. He adds that their luminosity shouldn't be different either for that kind of star so it must be accelerating away. Now you are in a fix! Do you assume therefore that he is right and space is expanding in an accelerating fashion? Otherwise how can you fit 116 bly into the space of only 100 Bly?

Because light from star 1 passed close to a number of gravitational bodies en route to you, it was bent by gravity one way and another. As I have indicated rather crudely, its *actual* distance covered is therefore the sum of all of its short journey legs; 116 Bly. This has knocked a bit of the stuffing (electrons) out of its photons and their red shift has measured their energy loss. When you looked at star 2, you were actually looking back along its last segment of the light's journey which is giving you an *apparent* source direction which is far from indicative of its *actual* source direction. So you have an *apparent* source position and an *apparent* *radial* distance when what you really need is an *actual* radial distance and an *actual* position.

To further complicate the issue, a deflection caused by a gravitational body will have a magnitude measurable by the angle of deviation. However, the direction the deviation then causes will depend on the lateral position of the gravitational body in relation to the light. This could be anywhere on a circumference of 360^0! Theoretically, various successive deviations could force light to follow extremely complicated routes en route including spiral paths.

Final Conclusions

- Space is neither curved nor expanding. Light is curved by gravity.

- Red shift is a measurement of energy lost by light en route from its source.

- Light is bent by gravity.

- Light will lose energy due to changes in its medium. Mediums include both encounters with gravitational bodies sufficiently close enough to affect its direction and that due to background cosmic radiation (gravitational red shift).

- A change in light's direction incurs an increase in journey distance and time.

- The total length of light's journey to an observer will rise exponentially in relation to the number and size of gravitational encounters and the length of time it is exposed to background radiation.

- Because the after-effects of gravitational encounters are exponential, red shift measurements will also increase exponentially.

- An exponential increase in distance creates an exponential increase in energy loss and an exponential increase in journey time.

- An exponential increase in journey time will cause an exponential increase in exposure time to both possible encounters with gravitational bodies and constant exposure to background radiation.

- Any of these effects will exponentially increase the *apparent* increase in an object's radial distance from an observer.

- These exponential effects have been largely the cause of the Big Bang theory, 'space expansion' and the 'Boson Particle'.

- When light's total distance travelled is increased, further opportunities of encountering gravitational bodies will occur.

- An observer has only the *current* means of measuring anything at all about the true nature of the Universe. It is red shift.

- This will relate to the total distance covered by light from its source. However, an observer has no means of determining an object's *actual* position. This will be distorted by the overall gravitational 'bending' of light on its journey to an observer.

- The direction light appears to come from (the image sighted) is not necessarily where an object is in the Universe. We will never know that since we have no history of its light's encounters en route.

- An observer has no knowledge of what may have happened to light from anywhere in the Universe during the time light has taken to reach him.

- Light from two objects which are *actually* equidistant from Earth but laterally separated by considerable distance could take different paths to reach a common observer.

- Light from the same source can take different routes to a common observer. Their respective distances covered en route would differ. Their red shift measurements would differ. Their positions in the sky could differ. An observer could wrongly 'ascertain' as a result of his observations that one was further distant than the other because it had accelerated faster and/or because space was expanding. An observer could be looking simultaneously at two slightly different views of the same object due to light setting out from it in two different directions. Twins?

- Space is not curved. 'There are no gravitational waves'. Gravitational effects can be varied in their intensity and direction whether curved, helical, spiral. straight, elliptical, circular etc. Light will be affected accordingly and its path will be modified to achieve equilibrium. These lengthened journeys will be longer and take more time, falsely interpreted as evidence of 'space expansion and acceleration'.

There is absolutely no way that red shift can be related to anything but the total *actual* distance covered by light from an object in space to an observer. This will include a relatively small Hubble Constant to account for gravitational red shift. By determining the amount of red shift caused by relatively close stellar objects, it should be possible to calculate a correction factor for much greater distances. However, despite being able to then interpret net red shift as a measure of *total distance covered* by light from a distant source, the number of deviations caused by gravity en route, its *actual* position and its *actual* radial distance from an observer will remain unknown. Our place in the Universe relative to everything there is in it will remain an illusion.

Basic trigonometry will demonstrate that any deflection of light however great or infinitesimally small will have an exponential effect on the *actual* distance it will have to cover to reach any prescribed radius from its source. Subsequent deviations however small will not 'balance out, detract from or otherwise affect the total distance covered to an observer. Consecutive variations in light's direction can cause light's route to be fashioned in many different ways; spirals, ellipses etc. These journeys may be complex or simple and impossible to either premeditate or calculate.

There are no easy answers to our conundrums! Returning to the analogy of my pilot, his examiner might have asked him whether he took any other factors into account on his journey such as the wind. The examiner might have explained that without the pilot being aware of the fact, he had encountered cross winds and head winds en route without knowing either their strength or direction. These effectively lengthened his journey by 18 miles since his course would have constituted a series of curves and not therefore, a straight line. This puts him in the position we find ourselves in now when we attempt to use the fuel consumption (red shift) of light travelling at a constant speed of 186,300 miles per hour from a distant star to measure its *actual* distance from us and its *actual* position.

I have a further illustration of how reaching conclusions with insufficient information is fatal.;

A pilot is asked as part of an examination to fly between two set map points A and B and measure as accurately as possible the distance between them. It's *actually* 100 miles but only his instructor knows that. The pilot knows he can maintain his aircraft's airspeed at 100 miles an hour (1.7 miles a minute) and also that flying at that speed its engine will consume 1 gallon of fuel a minute. It seems therefore a simple thing for him to fly from point A to point B as directly as possible, time the journey and by simple arithmetic calculate the distance separating them.

However, when he presents the answer to his examiner as 150 miles, did he consider how many deviations may have been caused by the wind on his journey? Was his altitude quite constant? Both of these would provide increases in *distance travelled*, fuel consumed and time of journey. If he relied only on his aircraft's fuel consumption (energy loss) to calculate the radial distance, it would be similar to the method we use armed only with red shift measurements. *Actual* distance covered would have been translated as *actual* radial distance. Let's hope he had enough fuel let over to land successfully!

The pilot will have had no precise knowledge of the strengths and directions of varying winds en route and how these affected his aircraft's fuel consumption and thus *apparent* radial distance. Scientists also appear to have largely overlooked that light reaching us from celestial bodies will have encountered an inestimable number of gravitational bodies en route which will have affected not only its total distance travelled, but also its red shift. The only concession they appear to have made is to propose a correction by means of the Hubble Constant.

Some afterthoughts

It's easy to see in retrospect how scientists have reached the conclusion thats space and the Universe are expanding. This was largely made possible because of the initial supposition 100 years ago that the Doppler effect (which implied that the recessional velocity of an object in outer space could be measured by the wavelength of sound emanating from it), could be applied to the wavelength of light emanating from objects in space (the amount of red shift). This denied the fact that so far we have found no justification that light can accelerate to over 186,300 miles per second!

It was established in 1919 that gravity could bend light so the Hubble Constant was established which attempted to introduce a correction factor to distances measured by red shift to take into account light's loss of energy due to gravitational effects. Telescopes weren't nearly as sophisticated and powerful as they are now and as objects at incredible distances came into our view, we were able to measure their *apparent* distances by red shift. In fact, *'Today, in the context of general relativity, velocity between distant objects depends on the choice of coordinates used and therefore, the red shift can equally be described as a Doppler shift or a cosmological shift (or gravitational) due to the expanding space, or some combination of the two.' (Wikipedia)*

Unfortunately for the protagonists of the Big Bang theory, these became more and more disproportionate to the *apparent* distances measured by other means even after taking the Hubble Constant into account. In my opinion, they incorrectly theorised that recessional rates were increasing with distance at an exponential (accelerating) rate. The rest is history.

We need to cease using measurements of light's red shift for any reason other than its total loss of energy en route to Earth. The Hubble Constant is a sensible means to apply a uniform correction factor due to the general gravitational force present in ex-galactic space. This is unlikely to cause very much of a deflection (bending) of light since gravity decreases with the square of its distance from a mass and there is a lot more space than matter in the Universe.

Since light can gain and then lose energy when being attracted towards gravitational bodies and subsequently lose energy when escaping from it, it seems that most of the problems reside in the fact that they do get bent in the process and that consumes time and distance in the process.

However, we are still left with an intangible problem for scientists to apply their wits to. We know that light can be affected when it passes close to large gravitational bodies including black holes but how can we legislate for the amounts of deviation these cause? Currently, we have no means of accurately measuring the distances of objects outside our immediate galactic neighbourhood. Parallax will do nothing but repeat the errors made by the deflections of light en route. I think I have provided enough proof that these effects are significant.

So, let's abandon chasing old theories with more and more unlikely, extravagant and expensive props to support them like 'expanding space' and Bosons which provide the energy of expansion'. I attempted to contradict these in my book 'Einstein's $E = mc^2$ unravelled – an alternative theory of the Cosmos'. I suggested that black holes are 'recycling centres' in which matter is converted into 'dark energy' with an infinite velocity. Strangely, it has recently been 'observed' that 'something is leaving a black hole at 99 times the speed of light'.

There most definitely is an *actuality* surrounding us. However, it most definitely is not currently accessible to us. Nothing we perceive is *'actually'* there and we merely fashion explanations from our perceptions, like imagining *reality* by interpreting it from the 'shadows on the walls inside Plato's cave'. As he suggested, we have to use our 'reason' to find even the grains of 'truth'.

Philosophy allows us to more easily imagine the Universe as being infinite and eternal with rules which maintain a constant entropic equilibrium. Our view of it is governed by the limitations imposed on us by the pedestrian speed of light and electromagnetism. We only have a perceived and distorted relationship with everything the Universe comprises, its planets, stars, galaxies and black holes because we cannot currently view them in *actuality*. Their substance is filtered through to us in unrelated scraps of information which we vainly attempt to interpolate into *reality*.

The Philosopher Leibnitz talked of a 'lattice' of monads; particles of matter so infinitesimally small they are invisible yet capable of interconnecting with everything there is and ever was in Singularity. Quantum physics describes electrons as being capable of 'communicating' instantly with one another by 'entanglement' across the farthest reaches of space.

I believe that the Universe is a stable entropic system in a state of Equilibrium and contrives always to correct aberrations; it's doing this right now because Homo sapiens have nearly wrecked Earth's prior stability. If there is a Universal Law of Equilibrium, then the Universe must surely need to be in direct communication with all of its parts. Quantum Theory may prove to be way it does so. Until we can use the instantaneous relationship between particles in which time and distance are absent, we will remain unable to discover the *actuality* that exists. We will remain blind to it.

We need huge quantum computers to do this for us large enough to contain everything there is. We do have a go at it however puny the effort. I can flit instantly from year to year or event to event across a span of 80 years in my own state of Singularity.

The Road-Map to Climate Safety

Preamble

I am truly fearful for my grandchildren whose futures are at risk not because we are unable to halt the process of Climate Change because, as you will see shortly, it is a completely feasible ambition. What worries me most is that governments including our own will fail to adopt a cohesive plan of action. They have offered us only a 'hotchpotch' of unconnected measures without the 'framework' of a rationally devised policy to hold everything together. Ironically, their pathetically small offerings appeal more to the major polluters of our planet, the coal, oil and gas barons. Their profits will I believe rise inexorably and their carboniferous products 'of every colour except green' will destroy Earth's species.

George Monbiot made the point in the Guardian of 20th October that we are avoiding the challenge of treating Climate Change as importantly as if we are 'going to war'. For example, our actions so far have been infinitesimal compared with the sacrifice, determination and commitment we were prepared to invest in defeating Hitler. The following day the Guardian reported that nations with huge reserves of fossil fuels were already displaying a lack of interest in COP26. It's not hard to see why; they want to protect the sources of energy which have made them rich. Coal, oil and natural gas were once essential to our growth and prosperity but now threaten our very existence.

Enter a new ally - Hydrogen

Hydrogen could help to snatch our salvation from the jaws of global political incompetence. Briefly, the sun delivers to the Earth more energy in just one hour than it consumes in a whole year! If we were to cover a total area of just 160,000 square miles in North Africa, Central America and Asia with solar panels, their sites would be almost invisible on a large-scale map, but could satisfy the world's entire energy requirements.

By deliberately generating surplus electricity, this could be employed to hydrolyse water into hydrogen to store and use to fuel gas turbines and balance supply and demand at night when the sun goes down. Sea water could be desalinated in North Africa and arid sites worldwide with access to the oceans, could be irrigated with desalinated sea water to develop new forests and replace the ones we have destroyed.

A year ago, I presented some practical suggestions to the then Secretary of State for Energy. The plans I put forward were economically attractive then. Now, huge price rises in the cost of electricity and natural gas have turned them from being interesting to imperative. Hundreds of other fine and necessary measures to locally tackle Climate Change have been made across the world. The BBC's Earthshot Awards described some but need the firm basis of an affordable and inexhaustible supply of green energy to structure them on. (See attached schematics).

Companies and countries with huge quantities of carbonaceous fuels such as coal, oil and natural gas are treating global warming as just another opportunity to make money. Already, they are putting forward ideas of how to produce 'blue hydrogen' using stocks of 'dirty' fossil fuels when green sustainable hydrogen can easily be produced from sunlight. My dread is that our politicians will continue deferring to the greed of carbon dioxide-emitting industries. Inviting their representatives to the COP26 conference is rather like 'giving arsonists boxes of matches to play with in a wooden house'. It would be wiser perhaps to treat them as our enemies.

The Australian Prime Minister claims that his country's coal industry sustains its economy but the truth is that nearly all the financial benefits accruing from the sale of their coal deposits go to that sector's owners, executives and shareholders. Solar panels installed across its sunny desert areas could transform the country. By powering desalination plants these could be turned into irrigated, fertile, agricultural land

and forest providing far more jobs than those currently deployed in its coal industry.

The ownership of UK's utilities passed into foreign hands with Margaret Thatcher. Their new owners are getting richer by the day threatening our very lives. We must create and own sources of clean energy quickly and take out the 'middlemen' whose energy supplies are not green in the true sense of the word. Only the tiniest proportion of the revenue accruing to all countries from the sale of fossil fuels benefits their economies and their workers who often epitomise 'slave labour'. Companies have been able to do this only by the political support they have engerdered and the financial power they can exercise. These corporate tyrants destroy forests turning hundreds of years' worth of carbon into wood pellets to burn in a mere day or two as biomass. What a misnomer that word is!

Not so long ago, £billions of inducements encouraged scientists to 'deny' that cigarettes smoking was harmful. A recent BBC documentary showed how 'climate denying' scientists have set back our global acceptance of Climate Change by over 10 years which may now threaten our existence.

Prince William is quoted as having said recently, "We need the world's greatest brains to work on Climate Change...." I disagree. We have all the necessary information on how to avoid catastrophe. What we really need is just plain common sense and some genuine unbiased cooperation.

Heat Pumps

- Recently, heat pumps have been seized upon by the Prime Minister as some sort of 'trailblazer' but in my opinion they fail both practically and financially. When we need to be reducing our dependency on imported energy by producing our own electricity cheaply and sustainably, heat pumps only increase our demand for it.

- When we should be looking at financially attractive means to reduce our carbon emissions, these will only increase our costs and subsequent demand for sustainable electricity.

- We need to search instead for ways in which to incorporate our assets, the natural gas and electric grid systems into an interlocking plan of action. Heat pumps will tend to make our gas grid redundant and will overload the electricity grid. By increasing our demand for electricity derived from fossil fuels, we are merely encouraging unsustainable energy production!

- Heat pumps offer no commercial or economic benefits and expose their users and the UK in general to excessive profiteering by energy suppliers. This is now clearly apparent with the exorbitant increases in the price of fuels derived in the main from fossil fuels of every description. The world is continuing to purchase electricity generated from coal for instance in USA, India, Australia, China and Japan. By buying energy overseas made by burning fossil fuels instead of generating it here, we make ourselves complicit in a mass betrayal of the Earth's species.

- Currently, to replace an existing domestic gas boiler with a net 45,000 btu/hour rated output, a heat pump would cost up to £18,000 and require up to 4.4 kW per hour to drive it costing a house-owner at today's energy prices up to about £1.10 per hour.

- This means that at peak demand, a heat pump could pull about 17 amps from the grid. In large numbers, this would place impossible demands on local sub-stations requiring significant upgrading of its infrastructure at a huge cost.

- A domestic heat pump will be of no help whatsoever when we all experience frequent spells of weather when the temperature exceeds 40 degrees Centigrade. A heat pump's function is to heat a house up, not to cool it down. They would still be obliged to install an air conditioning unit; even more cash outlay and a running cost of 75 pence to £1 an hour!

- Presently, I will go into detail about domestic solar panels. They could offer a solution by supplying 'home-grown energy' at an extremely low cost.

- Please note that I haven't approved the use of 'solar farms' as a source of the UK's sustainable energy production. These merely represent ways in which greedy energy companies can exploit us. Our government ceased providing tariff subsidies to home owners two years ago but are now open to giving huge sums of cash to subsidise companies which build solar farms and become today's new energy middlemen!

Cost of Imported Electricity

- From 'Statista', the UK imported a net 20,000 gigawatt hours of electricity in 2019.

- From 'Ofgem' the wholesale price of electricity has risen substantially since February 2021 from £53 to £79 per megawatt hour or £53,000 to £79,000 per gigawatt hour, a rise of no less than 50%.

- This has been due to no other reason than blatant corporate and national greed. It would be far more sensible to ration supplies of energy but it doesn't appear to be within Homo sapiens' nature to share anything with one another including food so no surprises there!

- Currently, very little of the world's production of electricity is derived from entirely 'green' fuel sources. They are mostly unsustainable sources of fuel and falsely described by their 'salesmen' as being 'sustainable' or in some cases 'blue'.

- Cutting down forests, turning trees into wood chips and burning it as 'biomass' is grossly misleading, scientifically inaccurate and insults what common sense we still have.

- Trees and many other organic species remain our principal means of extracting carbon dioxide from the atmosphere, emitting oxygen for us to breathe and slowly turning our 'arch-enemy' carbon into wood and eventually coal, oil and natural gas.

- Implying that trees and waste comprise healthy sources of energy by describing them as 'biomass' is quite misleading. The activity it promotes would be more accurately described as suicidal.

- In 2019, the UK's electricity sector's grid supply for the United Kingdom came from 43% fossil fuelled power (almost all from natural gas), 48.5% zero-carbon power (including 16.8%

nuclear power and 26.5% from wind, solar and hydroelectricity), and 8% was imported. Common sense tells us that our first target should be to replace the fossil fuel natural gas, with hydrogen. This would utilise our existing gas distribution infrastructure.

- The 'retail' price of electricity in the UK has risen this year from about 16 pence to at least 25 pence per kWh and its wholesale price is actually higher. This promises the lower paid serious financial difficulties. There is no tangible reason for this; it's entirely due to corporate and shareholder greed.

- This recent increase in the price of wholesale and thus 'retail' electricity will seriously affect any inducements to house and car owners to switch to using electric cars or heat pumps.

Some Home Truths

- Firstly, we need to find ways to tackle Climate Change in ways which will both reduce our import of electricity (which neither heat pumps nor electric cars will do). Solar power can do this, as I will explain, since it is freely available and cheap to 'download'. It can displace the 8% of our electricity needs we import from overseas which cannot be guaranteed to be 'green'. Most importantly, solar power can hydrolyse water into hydrogen giving us the most important advantage of all; flexibility. It can be stored to generate electricity as and when needed or used immediately in our natural gas grid system to reduce our dependence on fossil-fuelled energy.

- We need to set an example to the world at COP26. Whilst we remain willing to import or generate electricity from unsustainable sources derived from carboniferous fuels, we are defeating our own objectives. Let's take our 'battle stations' now and let our roofs provide the 'infantry' to lead our battle for survival.

- The average amount of carbon dioxide each private car pours into the atmosphere in the UK is 4.7 tonnes per annum. A published report states that by reducing the national speed limit to 50mph we could save about 15% of their fuel consumption. If therefore, the world's billion car owners could be gently persuaded to drive their vehicles at no more than 50mph and refrain from using their cars on just one working day a week, the world's total carbon dioxide emissions would be reduced by no less than 1.4 trillion tonnes per annum.

- When Paris choked on its emissions several years ago, its car-owners gladly agree to use them only on alternate days of the week so no problem there!

- Less servicing, less spare parts manufacture, less accidents, less pulmonary diseases, less hospital costs, less fuel transportation, less oil refining and less traffic holdups, would probably save another 15% taking the carbon emissions saved to at least 2.1 trillion tonnes. Another day leaving the car at home each week would take it to 2.8 trillion tonnes saved each year.

- Just these actions would save the world about £1.26 trillion on fuel costs alone. Surely, this could be the means to put Climate Change on the 'war footing' suggested by George Monbiot in this week's Guardian? How many coal mines would we close down?

The Undisputable Case for Domestic Solar Panels

- Domestic solar panels offer huge economic and geophysical advantages and represent the best way not only to meet our COP26 undertakings but also to boost our economy with the creation of hundreds of thousands of jobs financed by savings on our energy imports.

- In large numbers 4 kW domestic solar panel arrays could each cost £4,000 or less to install.

- On average, each 4 KW installation in the UK produces 3 megawatt hours per annum of electricity. Half of this, 1.5 megawatt hours, is normally consumed by the house owner leaving a surplus of 1.5 megawatt hours which he exports.

- Houses with roofs facing either East or West should also be considered for solar panels since their electricity outputs can be as much as 85% of those facing South.

- Energy companies currently pay solar panel house-owners a meagre 3-5 pence per kWh for their surplus production. My own domestic panels produce 4 megawatt hours per annum and I am paid only 4 pence per kWh for the 2.5 megawatt hours I export amounting to only about £100 per annum.

- If I need to buy back some of the electricity I have been exporting, I am currently billed by my electricity supplier about 14 pence a unit for it, a profit to him of about 250%. I even read the meters for him. When my fixed price contract expires In January, I expect him to charge his customers including me at least 25 pence per kWh for what we have to buy back from him giving him a profit of 525%. .

- Put simply, every house equipped with a 4 kW array could supply half of the needs of another house! Theoretically therefore, only half of the houses in an estate would need solar panels in order to supply half of the estate's total electricity requirements. That should interest Prince William whose Poundbury estate of around 2,000 houses display hardly any solar panels yet cost up to £1 million each to buy. Perhaps the Duchy could encourage just half of Poundbury's residents to invest only a fraction of what they paid for their properties and roughly speaking, generate half of their own electricity requirements and half of the remainder's. I understand one can now buy solar tiles in any colour now so this aesthetic consideration could well prove to be a deciding factor?

- Solar panels have a minimum life expectancy of 25 years and require little or no maintenance since the country's fairly large rainfall keeps them clean. An initial investment of just £4,000 would therefore provide an owner and his successors at least 75 megawatt hours of electricity worth £18,750 at today's retail cost.

- Domestic solar panel installations possess huge advantages over solar farms; they produce power 'to spare' at 'the point of use' avoiding the costs of adapting our grid infrastructure to cope with large localised inputs. This problem with wind farms. As another consequence, solar farms destroy agricultural jobs and new 'middlemen', landowners, reap the benefit of large lease payments. The result being that solar energy at an amortised cost of only 5 pence per kWh finishes up in our homes at 25 pence per kWh and rising.

- The recent price rises of electricity constitute blatant profiteering which has made investment in domestic solar panels overwhelmingly attractive. On average a householder would save about £450 p.a. on his electricity bills which, with his feed-in payments of about £60, would give him a return on his investment in only about 8 years and energy security for at least another 17. With global warming bringing higher temperatures, air conditioners will become a necessity. Because solar panels generate most of their output in summer, homeowners would have surplus energy with which to drive them at only about 4-5 pence per kWh, what he would lose by not exporting that amount of his surplus production. So, he could probably survive 40^{0}C temperatures during the day using completely green energy for only about 12-15 pence an hour or £1 a day!

- Electricity generated by domestic solar panels surplus to an owner's daytime requirements can be used to hydrolyse water locally to generate and store hydrogen to fuel gas turbines when the sun goes down.

- Our government wouldn't need to subsidise domestic solar panels. They are a 'snip' at £4,000 compared with to up to £18,000 for a heat pump and come with built in savings. Landlords and second-home owners could be charged with fitting them with solar panels at their own cost. We already pay them £23.4 billion a year via Housing Benefit so there shouldn't be too many complaints. New-build homes should be equipped with them as a planning requirement. To put things into perspective, we recently spent about £37 billion on 'track and trace' to protect us from Covid19 which threatens a relatively low percentage of our population with death and disease. Climate Change can represent a 'life sentence' to all of us. Can we please get our priorities right, Prime Ministers of the world?

- With some persuasion and a few 'sweeteners' house owners could meet the target of generating 20,000 (20 terawatt hours per annum) themselves wiping out the £1.6 bill on cost of importing the cost of our net import of electricity from dubious sources. Only 7 million domestic panel installations each generating 3 megawatt hours per annum would be required in our nation of 25 million homes to do so. That looks like an excellent means with car speed and use restrictions to 'quick start' a labour of love for this beautiful planet.

- At times when more power is produced than is required, surplus energy can be used to generate and store hydrogen at an on-cost of 35% of the production cost of solar power of 5 pence per kWh. 7.5 pence per kWh for hydrogen looks very attractive to me; only a fraction of the cost of imported unsustainable energy and comparable with cost of the fossil fuel, natural gas which provides almost half of our energy requirements. It's a 'no brainer'!

- Incidentally, the infrastructure needed to charge electric car batteries in sufficient quantities to justify them would be enormous. Many people cannot do so at home. However, those that could would be able to avoid the high cost of electricity at public charging stations if they recharged them during the day from their solar panels.

- I ask myself why solar panels aren't by now appearing on every roof from houses, factories, shops, office block, train stations, multi-stories and shop premises. I can only assume the obvious.

The Snake in the Cupboard

The biggest hurdle to overcome will be the fact that energy suppliers owned by overseas shareholders will, unless they are prevented from doing so, continue to hinder the essential move from fossil fuels to green sustainable energy. They act as little more than 'middlemen' in the passage of energy from source to consumer making huge profits of hundreds of percent. That can no longer be entertained. Taking a majority shareholding into public ownership once more must be considered.

In the Netherlands, water, electricity and gas networks are all publicly owned - and it's ILLEGAL to privatise any of them! That should tell us something.

Prologue

Hydrogen is the catalyst which can link every individual plan into a synchronised structure. Ideally, systems need to work in equilibrium and balance and only hydrogen will provide the flexibility and versatility to enable us to achieve that aim.

All of our ideas, imaginative views of a healed Earth or dreams of a world united in the pursuit of cooperation and equality, can only come about by recognising the value of hydrogen and its means of linking all of our truly environmental actions together. Like solar power, it is limitless and clean.

We can produce hydrogen from green power for only a relatively small on-cost making the total cost only about 7.5 pence per kWh. This seems cheap when compared to the 25 pence per kWh and still rising we are now paying for it. Solar power and hydrogen are 'gifts of nature'. We must fight fossil fuel companies tooth and nail if we are to use it. Their greedy ambitions currently threaten our very existence.

Hydrogen will facilitate ways to help balance our energy supply and demand, and importantly could still employ the infrastructures of natural gas and electricity grids our forefathers built and in which we have invested £trillions over the last 200 years.

Biography

Mike won a Grammar School scholarship in 1943 and after matriculating started work aged 16 as a trainee research chemist with the NCB studying chemistry at a technical college. After 2 years National Service, he worked in the Persian Gulf as a refinery operator followed by 9 years with Shell-BP and 2 years in Australia where he established a boiler manufacturing agency.

Subsequently, he worked for 6 years as a Senior Officer with British Gas responsible for the sale of natural gas to industry in the South of England.

After a short spell as a management consultant specialising in Human Relations, he became general manager of a company with 50 employees. He started his own business in water treatment in 1984 and invented a means to treat water without the need for chemicals which was granted a patent.

He has enjoyed many sports and leisure activities and represented Dorset County as a senior golfer. Aged 62, he trained as a pilot and qualified to fly in all weather conditions.

He has published two books of short stories and poetry and one entitled "Einstein's $E = mc^2$ unravelled".

This book challenges the Big Bang Theory and elaborates on the reasons why there wasn't one!

He suggests however, that the subject is completely irrelevant to our future on this planet and instead we should be devoting our energy and resources to fighting Climate Change!

9 781835 634097